I0494246

PLANT PROTECTION IN THE PACIFIC

SEMISI PONE
BSc, MSc (Hons), Auckland.

Copyright © Rainbow Enterprises 2013

Publisher: Rainbow Enterprises 2013

ISBN: 978-1-4993860-1-1

All rights reserved. No part of this publication maybe reproduced or transmitted in any form or by any means, electronic or mechanical, including photocopy, recording or any information storage or retrieval system, without permission in writing from the publisher and copyright holder. Rainbow Enterprises is the trade/publishing name of Semisi Pule aka Semisi Pule Pone.
Distributed by Rainbow Enterprises.
Second Edition. May 2014.
Cover photo – Vanilla Necrosis Potyvirus flexuous, filamentous particles. Photo by Professor M.N.Pearson, Biological Science, Auckland University.
Email: semisipone@yahoo.com

Notes on the Author: Semisi Pone attended Auckland University from 1981-1984 graduating with a Bachelor of Science in 1985. He started working for MAFF, Tonga in June 1985 and got a scholarship from the German Government in 1986 to complete his Master of Science degree at Auckland University, graduating in 1989. He continued working as a Plant Pathologist/Senior Plant Virologist for MAFF, Tonga until March, 1992 and moved to the University of the South Pacific, Agriculture Campus in Samoa where he was a Fellow in Tissue Culture for the European Union funded "Tissue Culture Project" from March 1992-May, 1993. He was appointed as the Plant Protection Advisor and Co-ordinator of the Plant Protection Service, South Pacific Commission in Suva, Fiji from May, 1993 to May 1996. He was also the Project Manager for the $US 3.5 million "Pacific Plant Protection Project" also funded by the European Union. He helped establish the Pacific Plant Protection Organisation amongst other work for the Pacific Region. He was also a member of the RPPO Technical Consultation and Biosecurity Expert Meetings at FAO, Rome. He has retired from Science and is now a writer of humour, children's stories, poetry, science and fiction (novels).

Note: All illustrations and photos were done by the author unless stated otherwise.

Introduction.

This book is a useful tool for reference to Plant Protection work that was carried out between 1985-1996 around the Pacific. It is based on my experience only, as a Plant Pathologist/Senior Plant Virologist for the Ministry of Agriculture, Fisheries and Forests in the Kingdom of Tonga (June,1985-March,1992), a Fellow at the Institute for Research, Extension and Training in Agriculture at the University of the South Pacific (March,1992-May,1993) in Samoa and Head of the Plant Protection Service/Plant Protection Advisor at the South Pacific Commission in Fiji (June,1993-May,1996). The South Pacific Commission is now known as the Secretariat for the Pacific Community with its headquarters in Noumea, New Caledonia.

I hope that this account of the work that we did, in the Pacific Islands, will be useful for current workers in National Plant Protection Organisations (NPPO) around the Pacific Region and RPPOs (Regional Plant Protection Organisations) in other regions of the world.

Many plant disease disasters have occurred in the Pacific Islands in the last 50 years which has shaped their current economies, social structure, political outlook and agricultural activities. It really shows how important and vital Agriculture is to the small island nations of the Pacific.

Banana Black Leaf Streak (*Mycosparella fijiensis* conidial state *Paracercospora fijiensis)* and Taro Leaf Blight (*Phytophthora colocasiae)* are the most obvious examples. BLS destroyed banana export from the Pacific Islands to New Zealand. At its peak, Tonga exported 20,000+ tonnes of bananas per year in the late 1960s to New Zealand. Fiji and Samoa were also supplying bananas to the New Zealand market at the time.

Taro Leaf Blight also destroyed the Samoan Taro Industry in 1993, which was its largest export at the time with about $ST10 million in earnings per year.

The losses due to these 2 plant diseases are more than a billion dollars in the 3 countries combined, since their introduction. This money would have been much needed in their economic development, education and community advancement.

It highlights the importance of Plant Protection in the Pacific as many plant diseases have not been introduced to the region yet. This includes viruses of cassava and papaya. Any new disease attacking crops in the Pacific region will have the same devastating effect as Banana Black Leaf Streak and Taro Leaf Blight.

Agricultural exports from Samoa and Tonga are still minimal, with less than 30x20 tonne containers of rootcrops exported per year (FAO Statistics, 2009). Only Fiji is doing well in its agricultural exports with more than 600x20 tonne containers of rootcrops exported per year, probably to Australia and New Zealand (FAO Statistics, 2009).

All the other 19 Pacific Island territories and States are also in the same category as Tonga and Samoa with minimal or no export of rootcrops probably due to many constraints including plant diseases.

This book is not intended as a textbook but an account of the work I was involved in around the Pacific from June, 1985 – May, 1996. I have ensured all the information are correct in case readers may want to refer to or quote from this book. References are quoted, otherwise the information is derived mostly from my own experiences and personal communications.

Further readings are listed at the end.

CONTENTS

CHAPTER 1. BANANA PRODUCTION IN TONGA.

Banana export from Tonga was second only to copra export as the main foreign exchange earning for the country prior to the introduction of Black Leaf Streak in the 1960s. It is thought to have been introduced from Fiji where it was known as Black Sigatoka because it was first discovered in the Sigatoka Valley.

Scientists at the time thought it was a new strain due to its virulence, hence the name "Black Sigatoka". A description of the black appearance of the banana leaves when destroyed by the disease. Banana production and export from Tonga was at its peak at around 1968 when more than 20,000 tonnes were exported to New Zealand in that year.

BLS was introduced a few years earlier and it soon overwhelmed the banana monoculture plantations at that time. Despite the introduction of chemicals to control the disease, production crashed to as low as 500 tonnes exported per year, a decade later, in the 1980s.

A $5 million banana rescue project was established by Tonga MAFF and the New Zealand Government but it did not save the Banana Industry. By 1990 it had completely collapsed.

Banana pests simply made it too uneconomical to grow and export bananas from Tonga and its neighbours (Fiji and Samoa). Although banana is still grown in the Pacific in "mixed cropping situations", with no chemical input, it is no longer a viable export option unless costs can be reduced.

The losses to the Tongan economy is probably around $TOP400 million or more, in terms of lost revenue (assuming $TOP12 per 20kg carton) since the introduction of BLS. It is probably around the same or

Note: Taro Leaf Blight has cost Samoa about $ST 200-300 million in lost exports since it was introduced in 1993. The loss to the local market is much bigger. Although, taro can be grown with chemical control, most of the 150,000 people in Samoa do not have regular supplies like they used to. Taro in American Samoa was also destroyed by TLB in the same epidemic. Fiji now supply more than 90% of the taro in the Australian and New Zealand markets, with new players like Vietnam supplying some shops.

higher in Fiji and Samoa.

Bananas were traditionally grown in mixed traditional cropping systems in Tonga and around the Pacific Islands. It also included plantain and lesser varieties.

Monocultures were introduced in the 1960s as a response to the increasing demand from New Zealand for ripe bananas from the Pacific. No one else was supplying it.

Ecuador and the Philippines later sold their bananas in New Zealand and slowly took over the market. Probably due to the production and disease problems in the Pacific but also due to better presentation, a cheaper product and economy of scale. The Pacific Islands simply cannot compete with these larger countries.

Despite the claims of better taste from the Pacific Islanders better appearance finally won the race. Postharvest diseases and packaging problems plaguing Pacific Island banana export cannot be controlled and contributed to the final demise of the Banana Industry and its exports from the Pacific.

Ecuadorean bananas simply look better in the shops and supermarkets of New Zealand, with no marks on their beautifully presented, yellow, mouth watering bananas.

On closer examination it became obvious that Banana Production in Tonga has too many problems to deal with, most of all the pests. I have listed the limitations to banana export production, in Tonga, to demonstrate this point. It also highlights the problem faced by export banana growers from the Pacific Islands.

TONGAN AGRICULTURE...

Tongan traditional agriculture was based on "cut and burn" practices. The forest trees are cut and burned, then crops are planted. Yams are the first to be planted then followed by taro, then cassava or sweet potato or vegetables. After several crops, the piece of land is left to "fallow" or regenerate for several years. The farmer moves on to a new area and so on. The forest trees grow back and then crops are planted again. Traditional crops rotation and mixed cropping were later found to be effective in reducing disease and insect pest incidence. However, due to the rapid increase in population and pressure on limited land, fallowing is reduced to just a few months or a year. Long term crops are also used as fallow instead of grass and trees. Cassava and taro take up to 8 months to mature so are excellent substitutes.

LIMITATIONS TO BANANA PRODUCTION IN TONGA.

1. LAND PREPARATION

Prior to monoculturing, bananas were grown in mixed crops with yams, taro and other rootcrops. The monoculturing practice was introduced to increase productivity from smaller areas of land. Instead of 20 banana plants per acre, for example, it became 500 plants or so per acre. Everything changed. The production practices have to keep up with the new intensive cultures. Tractors were introduced to till the land before planting. Banana growers have to pay for the land preparation and it became a major cost of establishing a banana plantation. Before the tractor was introduced, a grower can just use his digging spade or fork to plant them, now he has to use a tractor to prepare his piece of land before planting.

2. PLANTING MATERIAL AVAILABILITY

Finding large numbers of sword suckers for planting a new plantation was a challenge. Most mixed plantations did not have enough.

Sometimes nurseries have to be established to provide them. The suckers have to be cleaned of excess soil and roots to avoid introducing banana nematodes to the new land where the plantation will be. Experience suggest that the plantation will need no nematode treatment for the first few years, if the planting material is clean.

3. FERTILISER COST

Prior to monoculture practices there was no need to fertilise. The old method of fallowing was used in mixed cropping. Intensive planting of several thousand banana plants in a plantation, which may be harvested for several years necessitates fertilising or the production will not be profitable as bunches decline in size and weight over the years.

4. EQUIPMENT COST

When Black Leaf Streak was introduced, mistblowers and protective uniforms, were required for fungicide sprays. Together with the chemicals, they were major costs. The scabmoth control method also needed a special injector. Chemical weed control needed knapsack sprayers. Desuckering also needed another piece of specialised equipment. They all add up.

5. CHEMICAL COSTS

Chemicals are also a major cost in banana production. Fungicides for BLS control. Insecticides for scab moth and aphid control. Nematicides for nematode control and also post-harvest chemicals.

6. LABOUR COSTS

Planting, weeding, harvesting and packing needed a large amount of labour. An eight acre plot of bananas may have up to fifty or more bunches to be harvested every two weeks. Growers need some extra hands to do it quickly before the boat leaves.

7. WEED CONTROL

In large plantations of more than 20 acres, for example, weed control is a major cost. Several labourers with knapsacks are required to spray weedicides every 2-4 weeks. The old method of weeding with hand held hoes are too slow and even more time consuming and costly.

8. NEMATODE CONTROL

As mentioned before, if the planting material is clean, growers can get away with no nematode control for 2 years or so before they need to. Heat treatment of planting material was even tested for elimination of nematodes, but was not feasible in my opinion.

Nematode chemicals, like the other chemicals, were also very costly.

9. BANANA BUNCHY TOP VIRUS (BBTV) CONTROL

Clean planting material and a clean area are the first steps to controlling BBTV. Alternative hosts like wild ginger should be removed from the immediate vicinity. The second is aphid control. BBTV is spread by the aphid *Pentalonia nigronervosa* through feeding on diseased banana plants then transferring the virus particles to healthy plants. The virus remains in the aphid for a long time, after feeding on infected plants; perhaps up to 2 weeks. When aphid populations on suckers reach a certain number, the adults develop wings and fly off, probing on other banana plants thus

Figure 1. Banana Bunchy Top Virus infected banana sucker. Note the upright "bunched" leaves.

14

spreading the disease. All banana stems and suckers are sprayed with aphidicides to reduce their numbers and prevent them from developing wings and migrating. Removal of infected banana mats and replacing them

with clean suckers also helped. Healthy suckers are easily recognised because of their vigorous and "healthy" appearance. They should be selected from a healthy plantation were BBTV is not present.

Figure 2. Healthy banana sucker suitable for planting.

BBTV, like Black Sigatoka, first appeared in Fiji. Then it was reported from other parts of the world. It can be a serious problem in plantations because of the long delay before harvest, after the infected plants are removed, and new healthy suckers planted. This affects the number of bunches harvested and cartons exported. BBTV is usually not a problem in well managed plantations.

10. BLACK LEAF STREAK (BLS) CONTROL

In large plantations of more than 20 acres, this is probably the most expensive of all banana production costs. BLS fungicides have to be sprayed every two weeks. Labourers with mistblowers spray the fungicide into the air between banana rows. The idea is to cover the leaves with a fine mist of the chemical. Misting oil and a "sticker" is added to help stick the chemical to the banana leaf and

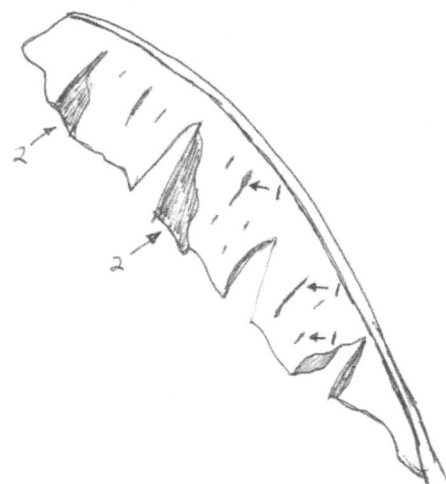

Figure 3. Banana leaf infected with Black Leaf Streak. Note the black streaks (1) and the spreading necrotic areas (2).

"mists" and settle the chemical quickly so there is no drift. If it is too windy, chemical drift may render the practice useless because the chemical "mists" may be blown out of the plantation.

Black leaf streak can be minimised through cultural practises such as crop rotation, mixed cropping and the use of barriers.

11. HURRICANES

Hurricanes and strong winds are very frequent in the Pacific Islands from February to March. Sometimes earlier or late in the year.

When strong winds arrive, it is usually 100% loss of all banana plants in the plantation. There will be no harvest for another 8 months or so as new suckers grow and become productive.

This is a massive loss and sometimes kills the whole operation or adds to the costs.

12. DROUGHTS

Dry periods are frequent during the middle of the year. Sometimes as long as 3 months, but that is very rare. Usually a few weeks, but it may be enough to lower production as bananas need a lot of water. It also affects nematicide and fertiliser effectiveness as they need water to carry them down to the banana roots.

13. HANDLING AND POSTHARVEST PROBLEMS

One of the biggest problems for Pacific Island bananas in the New Zealand market, over the years, were the handling and postharvest diseases like anthracnose. Black marks or bruising on the fruits due to the handling problems made the fruits unattractive to potential customers. The Tongan and Pacific Island banana suppliers never really figured out a way to avoid it and in the end lost the market to South American and Asian suppliers.

14. PACKAGING COST

Tongan bananas were usually packed in plastic bags inside cardboard cartons. The cut ends of the banana hands were also treated with a fungicide. Although packaging was not considered a major cost, it still added to the total, but a necessary one.

15. TRANSPORT AND SHIPPING COSTS

Most smallholders used horse drawn carts or small 1-2 tonne "pickups" to harvest and carry their banana bunches to the packing shed. In most cases, a distance of several kilometres. Roads were full of potholes and the constant rocking of the carts and vehicles bruised the banana fingers as they rub against each other and the floor of the cart. Sometimes growers use old banana leaves to pad the floor to avoid bruising.

Packing in the plantations were encouraged to avoid bruising in the later years. Even chemical dipping before export was used to reduce bruising and control anthracnose. Shipping is also part of the cost growers pay for, but that cost is taken off before they get paid.

16. PRESENTATION AT MARKET.

After all the production and work done if the product is not attractive to the customers, they will not buy it. That is exactly what happened to the Tongan and Pacific Island bananas when the Ecuadorean and Philippine bananas were also available to New Zealand customers, and the cause of the final demise of the banana industry in the Pacific Islands.

17. HIGH PRICE/PROFITABILITY

Banana production was very profitable to many planters in Tonga, Samoa and Fiji despite the intensive practices and costs. Even when the market price was as low as $TOP 6.00 per 20 kg carton, it was still profitable due to high yields/production.

A POSSIBLE SOLUTION TO DISEASE PROBLEMS IS TISSUE CULTURE.

The advantages of Tissue Culture include (i) rapid multiplication of disease free popular crops (ii) ease of transferring large quantities of plantlets between countries (iii) restarting plantations with clean plantlets (iv) storage of germplasm for future generations and research.

Figure 4. Tissue cultured vanilla plantlets in bottles. Vaini Research Station.

Banana plantations in the ROC, Taiwan are replanted with clean planting material after every harvest. This practice avoids diseases like BBTV and minimises scabmoth and BBLS. Production is controlled and target volumes are reached on time.

Figure 5. Banana plantlets, from Tissue Culture, being watered in the "Greenhouse". Banana Research Institute, Republic of China on Taiwan.

One of the problems facing banana production in the ROC, Taiwan is bacterial rot caused by *Pseudomonas* spp. The infected plants are removed and replanted with new clean material (foreground). *Pseudomonas* is usually a problem in waterlogged areas. Perhaps the irrigation system used in the ROC is the problem. When bacteria gets into it.

Figure 6. Tissue cultured banana plants growing vigorously in the field. Banana Research Institute, ROC, Taiwan.

CHAPTER 2. VANILLA NECROSIS POTYVIRUS (VNPV).

Vanilla export is a major foreign exchange earner for the Kingdom of Tonga. It may be earning as much as $TOP 5 million in the "good years". Sometimes it is difficult to control production as many growers do not manage their plantations properly and sometimes plantations are turned into a yam plot or another more profitable crop when vanilla prices fall. However, like most economic products, prices will change due to supply and demand. In the good years, a kilo of dried vanilla beans can fetch up to $US132.00. It has been the cause of much discussion between the Ministry of Agriculture, Fisheries and Forests and growers in Tonga. They need to save money in the good years to help in the bad years.

Figure 7. Vanilla bean production on healthy plants can be maximised by good management.

The discovery of the viruses affecting vanilla in Tonga was of much concern at the time (1986), because vanilla was very popular with farmers and new plantations were being established at a very rapid rate through the MAFF Vanilla Development Project. A joint research project between the University of Auckland and the MAFF Research Division identified the viruses as ORSV, CyMV; two common viruses of orchids and VNPV a previously undescribed virus.

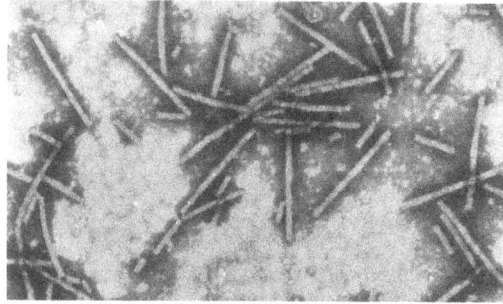

Figure 8. Odontoglossum Ringspot Virus (ORSV) (x 60,000) appears like "match sticks" under high electron microscope magnification.

Both ORSV and CyMV are found in ornamental orchids all over the world. These orchids are also grown and sold commercially.

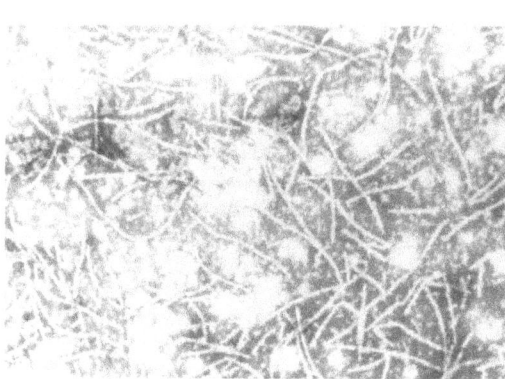

Figure 9. Cymbidium Mosaic Virus (CyMV) (x 60,000) appears like "cotton wool" filaments under high electron microscope magnification.

Most commercial operators do not test for or worry about ORSV and CyMV in their orchids. It was only recently that scientists have tried to clean commercially grown orchids of these viruses through Tissue Culture.

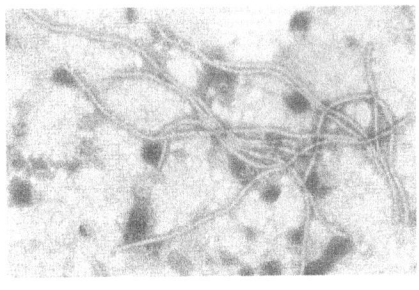

However, in the case of VNPV, our research have found it to be very destructive to vanilla plants. Young shoots and leaves become necrotic and die.

Figure 10. Vanilla Necrosis Potyvirus (VNPV) (x 60,000 approx.) appears like filamentous, flexuous "spaghetti" under high electron microscope magnification. These VNPV particles were from *Nicotiana benthamiana* leaf tissue. Photo by Professor MNPearson, University of Auckland.

Newly infected plants dies back from the tip and shows chlorotic sunken areas on older leaves, which turn necrotic and die as the disease progress. It rapidly loses

all its leaves leaving only the vine, which sometimes regrows and "dies back" many times before the whole plant dies.

Figure 11. Young vanilla shoots "dieback" from the tip and leaves become chlorotic then necrotic.

In new plantations the disease can be spread very quickly at this stage especially if there are herbaceous weeds in and around the plantation which hosts large numbers of aphids. Aphids fly around the plantation, probing on diseased plants then healthy plants thus transferring the disease.

Laboratory studies have shown that VNPV can infect herbaceous weed hosts. Indicator plants like *Nicotiana benthamiana* and *Nicotiana clevelandii* are infected systemically. Symptoms include chlorotic veins and distorted leaves. *Chenopodium amaranticolor* and *Chenopodium quinoa* develop pinpoint yellow cholorotic spots on their leaves when infected with VNPV (see section on alternative hosts).

FIRST SYMPTOMS FOUND.

Symptoms of the Vanilla Necrosis Potyvirus (VNPV) were first noticed in the research vanilla plot at the MAFF, Vaini Research Station. Samples of the leaves with sunken chlorotic, areas and distortion were sent to Dr Michael N Pearson of the University of Auckland (1986) who examined the sap under the electron microscope and found "potyvirus like" flexuous, filamentous particles in it. Potyvirus like particles was also reported by Mossop et al (1984) in earlier surveys. A potyvirus was also reported to affect *Vanilla tahitiensis* in French Polynesia by Wisler et al (1987). It was later found to be unrelated to VNPV.

PARTICLE LENGTH MEASUREMENTS.

One of the determining factors for potyvirus diagnosis
is the particle length. They usually fall within 680-900
nanometres. Extracts of partially purified preparations
of VNPV from symptomatic vanilla leaves were
photographed under the electron microscope then the
particle lengths measured.

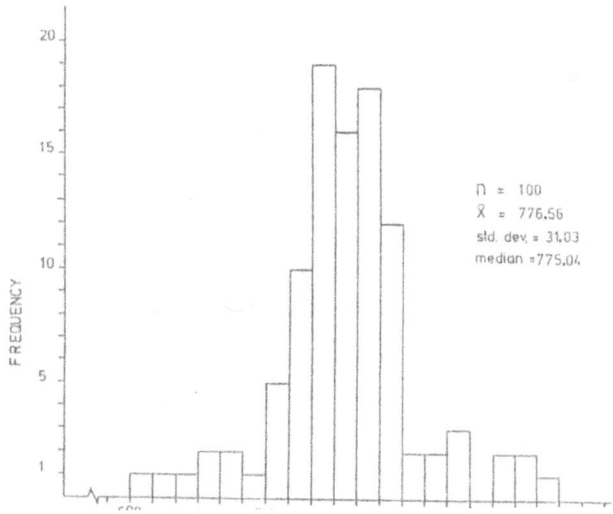

**Figure 12. Particle length measurements shown in a bargraph. Average
length was 776 nanometres.**

Particle lengths were within the potyvirus group with
an average length of 822.88 nanometres. Particle
lengths of preparations from vanilla sap were also

measured and were within the potyvirus particle length average length = 776 nanometres, Figure 12.

TEST OF CUTTING GROWTH

Healthy and VNPV infected vanilla cuttings were planted in plastic bags in the glasshouse. Their growth were observed. After 3 months the VNPV infected cuttings had developed shoots but were stunted. They were less than 30 centimetres in height. Healthy cuttings were about a metre high. A difference of about 70 centimetres in growth. Infected cuttings planted in the field are also stunted and never grow more than a metre high before leaves and the stem become necrotic and die. Growers were advised to choose only healthy, non-symptomatic plants as their source of planting material for new vanilla plantations. It was a simple but very effective way of excluding the VNPV from their new plantations, effectively controlling the disease.

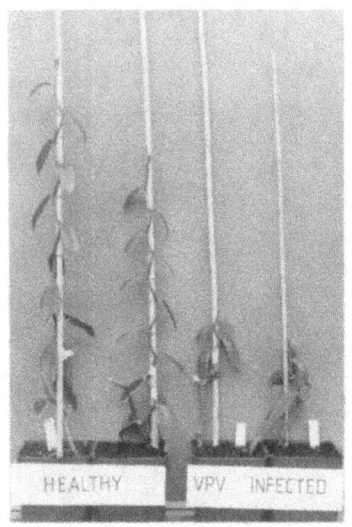

Figure 13. Healthy and VNPV infected vanilla cuttings were planted in the lab. After 3 months the healthy plants were 1 metre high while the infected shoots were stunted.

FIELD CONTROL OF VNPV.

In combination with aphid control through removal of herbaceous hosts from the plantation, using selection of healthy plantations and planting material was very effective in reducing the VNPV incidence in vanilla plantations in the whole Kingdom. Plantations throughout the Kingdom of Tonga were surveyed and infected symptomatic plants were removed and replaced with healthy planting material.

The total of exported dried vanilla beans from the Kingdom jumped from 20 to 70 tonnes some 3 years later. It was probably a combination of the effective virus control, better management and the increase in acreage due to the Vanilla Development Project.

> Note: It has been reported in the literature that similar symptoms to VNPV had appeared and destroyed the Vanilla Industry in Bali, Indonesia. Production fell from 4093 hectares producing about 279.7 tons of dried beans per year to only 5.5 tons of dried beans from 360 hectares per year. Most plantation vanilla plants showed shoot and stem necrosis, followed by rapid loss of leaves and collapse of the plants. Stem rot caused by *Fusarium oxysporum f.sp. vanillae* was quoted as the cause of the disease. Those symptoms are, usually, what happens with vanilla plants in Tonga that were newly infected with VNPV (SPPone, 1988). An ELISA test with VNPV antiserum should prove the presence or absence of the virus.

Vanilla plants infected with VNPV can be clearly identified through symptoms only. Growers were advised to remove infected plants when symptoms appear.

Figure 14. A vanilla plant infected with VNPV loses all its leaves very quickly. Vaini Research Station, MAFF, Tonga.

ORSV and CyMV are two common orchid viruses and are usually present in most orchid species with no apparent harm done to the plants, although mottles and

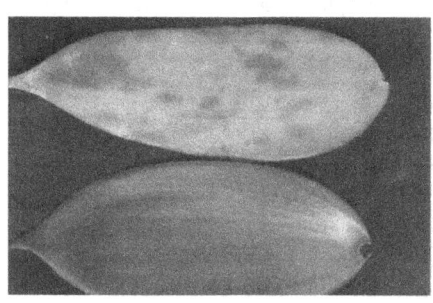

isolated necrotic spots may appear on orchid leaves. Leaf mottles are also associated with vanilla plants infected with ORSV and CyMV but plants are still productive, although

Figure 15. Vanilla leaf showing mottling). there may be a **symptoms (top) and healthy vanilla leaf (bottom)** stunting effect on growth which needs further investigation (Pone SP, 1988). Mottling does not appear to have any adverse effects on vanilla plants, affected plants should not be used as a source of planting material.

ALTERNATIVE HOSTS OF VNPV.

VNPV can also infect other species of plants in laboratory studies. It causes "pinpoint" chlorotic spots on *Chenopodium amaranticolor* and chlorosis which turns necrotic on *Nicotiana clevelandii*.

It is possible that alternative hosts are also present in and around vanilla plantations in Tonga. This may be why rapid spread of VNPV occurs in some plantations.

Figure 16. "Pinpoint" chlorotic spots on *Chenopodium amaranticolor* leaves caused by VNPV.

Nicotiana clevelandii leaves first appear chlorotic then becomes necrotic around the edges (see Figure 17). It is the same symptoms in vanilla

Figure 17. A *Nicotiana clevelandii* leaf showing diffuse chlorotic areas which later turns necrotic (left). It is the effect of VNPV when "mechanically" inoculated into the leaf. The leaf on the right is a healthy leaf.

where the young leaves become chlorotic then necrotic. It emphasises the destructive nature of VNPV on infected plants. Alternative herbaceous hosts of VNPV in vanilla plantations should be identified as part of control strategies in large vanilla producing countries where VNPV is present. If the inoculum and vector is removed from the field, effective control is achieved. This strategy can also be used for kava and other virus diseases of agricultural crops.

VNPV infected *Nicotiana benthamiana* leaves also produce veinal chlorosis. A large amount of VNPV was purified from *N. benthamiana* leaves

Figure 18. *Nicotiana benthamiana* leaf showing faint chlorotic veins (A) compared to healthy leaf (B).

(See Figure 10).
VNPV particles were also partially purified from symptomatic vanilla leaves for the development of the antiserum for the ELISA test.

THE DOUBLE ANTIBODY SANDWICH - ENZYME LINKED IMMUNO-SORBENT ASSAY(DAS-ELISA) PRINCIPLES (Clark and Adams, 1977).

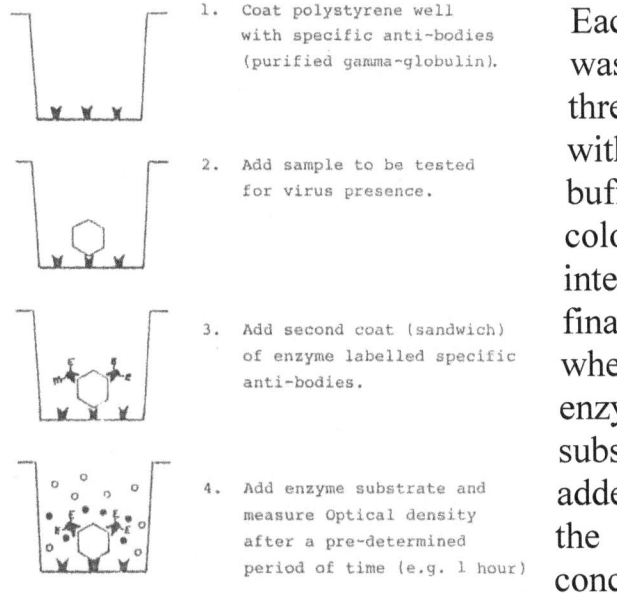

1. Coat polystyrene well with specific anti-bodies (purified gamma-globulin).

2. Add sample to be tested for virus presence.

3. Add second coat (sandwich) of enzyme labelled specific anti-bodies.

4. Add enzyme substrate and measure Optical density after a pre-determined period of time (e.g. 1 hour)

Each step is washed out three times with a special buffer. The colour intensity of the final reaction when the enzyme substrate is added indicates the concentration

Figure 19. The principles of the ELISA test.

of the virus in the sample. In the case of VNPV, the enzyme was alkaline phosphatase which results in a yellow colour. The reaction of VNPV infected plants in the field in Tonga was very strong. A potyvirus group test from Sigma USA was also used and the absorbance (optical density at 405nm) were very similar at 0.200-0.500 after 30 minutes.

DEVELOPING THE ANTISERUM.

Partially purified VNPV particles from frozen symptomatic vanilla leaves from Tonga were injected into a rabbit 3 times. Sub-cutaneously and intramuscularly. The rabbit was bled after 21 days and the serum separated from the red blood cells. Healthy antigenic proteins were absorbed, using healthy vanilla tissue, and the VNPV antibodies was removed from the serum. The ELISA test was developed using the VNPV antibodies (Figure 19) and optimised in the University of Auckland Biological Science Laboratory for use in the MAFF Research Laboratory in Tonga. The antibody was conjugated with the enzyme alkaline phosphatase. A yellow colour develops when the substrate is added. The yellow colour and absorbance are indicative of the virus concentration of the sample. The complete process of purifying VNPV, producing the antibodies to VNPV and developing the ELISA test for VNPV is explained in **"Developing an ELISA Test for VNPV in the Kingdom of Tonga"** by Semisi Pone. This book is available from Rainbow Enterprises. Information on developing the ELISA Test is also available in **"An investigation of 3 virus diseases of *Vanilla fragrans* (Salisb) Ames in the Kingdom of Tonga"**, MSc Thesis (1988). S.P.Pone,

Biological Science Library, University of Auckland, New Zealand.

TESTS OF PATHOGENITY.

VNPV was associated with all plants that showed symptoms of dieback, necrosis and loss of leaves. Tests were also carried out to prove whether VNPV was the cause of the symptoms. VNPV infected tissue was used to mechanically infect healthy vanilla plantlets in the laboratory by rubbing the infected tissue into the vanilla leaf and grafting infected tissue into the stem to try and "recreate" the disease symptoms in the new shoots.

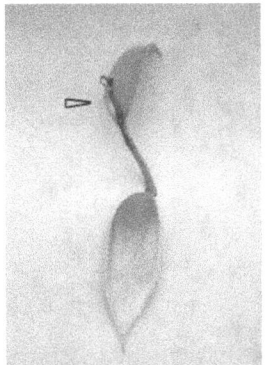

One of the most important steps in epidemiology studies of a plant virus disease is to prove the disease symptoms is caused by the virus. Mechanically infecting healthy plants is the best way to do it. The symptoms will then be used as indicators of disease presence in the field.

Figure 20. Tip "dieback" of vanilla plantlets in the lab after being infected with the VNPV a few weeks before.

Disease symptoms that are unique to the particular virus being studied is a very effective and "easy" way of identifying infected plants in the field. In the case of vanilla where there are thousands of plants in one plantation, the symptoms was very effective in identifying and studying virus epidemiology. It was clear that the virus was spreading rapidly in plantations with a large number of aphids present. In 3 months about a quarter of the plants were infected, which started with a few plants. Spatial analysis conclusively proved that spread of the virus is not random but clumped, contagious or clustered. This means that the probability of a healthy plant being infected is very high if it is next to an infected plant (P=0.025). P is the significance level.

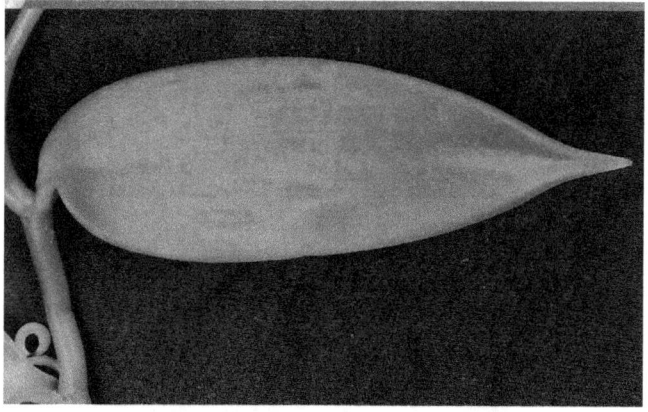

Figure 21. Diffuse chlorotic areas on mechanically infected young vanilla leaves, in the laboratory. The symptoms are very similar to that caused by the virus in the field. See Figure 7.

In the case of VNPV it was proved, mathematically, to be spread by growers during practices of looping, flower initiation and pollination along the rows and aphids across the rows. Tests for clumping were significant at P=0.025, which means that the distribution of the infected plants are not random but are influenced by aphid migration and probing within the plantation. A complete discussion of the mathematics and epidemiology of VNPV can be found in **"Epidemiology of Vanilla Necrosis Potyvirus in *Vanilla fragrans* (Salisb.)Ames plantations in the Kindom of Tonga"** by Semisi Pone. Copies can be obtained from Rainbow Enterprises.

GRAFT INFECTION TEST

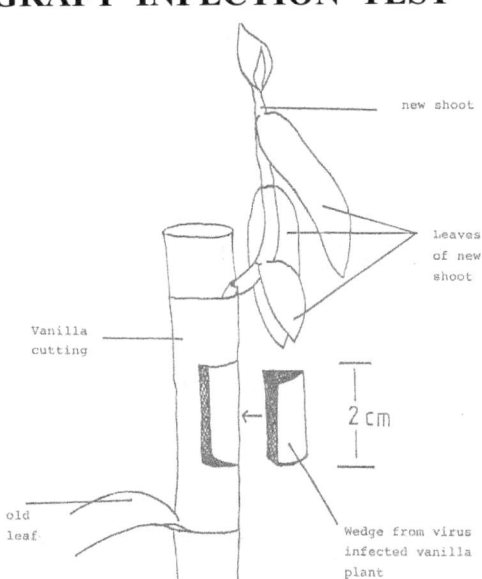

new shoot

Leaves of new shoot

Vanilla cutting

2 cm

old leaf

Wedge from virus infected vanilla plant

Figure 22. Scheme for grafting infected vanilla tissue into healthy plants to recreate the disease symptoms in the laboratory.

The results of VNPV, CyMV and ORSV infected wedges grafted into

35

young vanilla plantlets singly and in combinations were not satisfactory. Failure of vascular tissue to join as well as non-transmission of viruses through parenchyma tissue may be the reason. In other plant species viruses have been reported to be transmitted quickly through the connection of vascular tissue.

POTYVIRUS EVIDENCE.

Viruses in the Potyvirus Group have filamentous, flexuous particles that are between 680-900 nanometres long (see Figure 10). They also cause "pinwheel" inclusions in the cells of plants they infect.

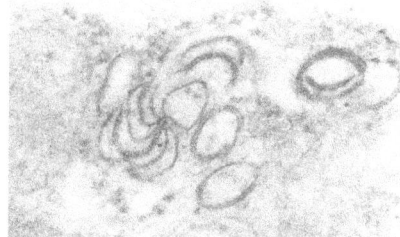 These properties lead to the conclusion that VNPV is a potyvirus. A member of the most destructive group of plant viruses.

Figure 23. Pinwheel inclusions caused by VNPV in _Nicotiana benthamiana_ cell tissue. Photo by Professor MNPearson. University of Auckland.

Losses due to the Potyvirus Group members are probably more than any other group in terms of economic value to agriculture worldwide. Many of the most important viruses that cause huge losses in economic crops worldwide are members of the Potyvirus Group. Example are Potato Y viruses and

Zucchini Yellow Mosaic Virus (ZYMV) which attacks cucurbits.

Potyvirus tests usually include (i) particle measurements (ii) presence of pinwheels (iii) aphid transmission (iv) serological/ELISA tests. All these conditions have been satisfied in the case of VNPV.

APHID TRANSMISSION

Professor MN Pearson of the University of Auckland carried out some studies of VNPV aphid transmission between *Nicotiana benthamiana* plants, during his sabbatical in the United Kingdom, and found that aphids *Myzus persicae* and *Aphis gosypii* can transmit VNPV between *Nicotiana benthamiana* plants.

I also carried out some aphid transmission tests on vanilla plantlets in Tonga at the MAFF Research Station using *Aphis gossypii* but no symptoms were observed on the vanilla plants after 3 months. According to Professor Zettler et al, they were able to transmit a potyvirus of *Vanilla tahitiensis*, from French Polynesia, using aphids at the University of Florida but it took up to 8 months for symptoms to appear. Perhaps my tests in Tonga needed more time before symptoms of VNPV appear.

Aphid transmission of potyviruses are non-persistent. This means that the virus is only infectious for a brief period, perhaps a few minutes. Persistent viruses can be transmitted by the aphid for a long time after being picked up through probing on infected plants, probably several hours or days. Non-persistence transmission can be used in designing control measures which aim at delaying aphid transmission between plants.

Studies of aphids and ZYMV infecting commercial squash plantations in Tonga have shown that aphid migration within plantations may be affected by wind direction. No observable effect was found in vanilla plantations.

APHIDS FOUND IN TONGA.

Dingley et al (1981) listed 8 species of aphids found in Tonga. The most common were the grass aphids (*Rhopalosiphum maidis*) mostly found on Johnson grass and the green aphid (*Aphis gossypii*) which populate herbaceous weeds in plantations as well as most crops and plants. *Aphis gossypii* is probably the most important in terms of virus transmission, of all virus transmitted viruses, between plants and also between plantations. In a study of aphid behaviour and the transmission of ZYMV, in squash plantations in Tonga, Aphis gossypii was the most commonly caught in the aphid traps.

EPIDEMIOLOGY

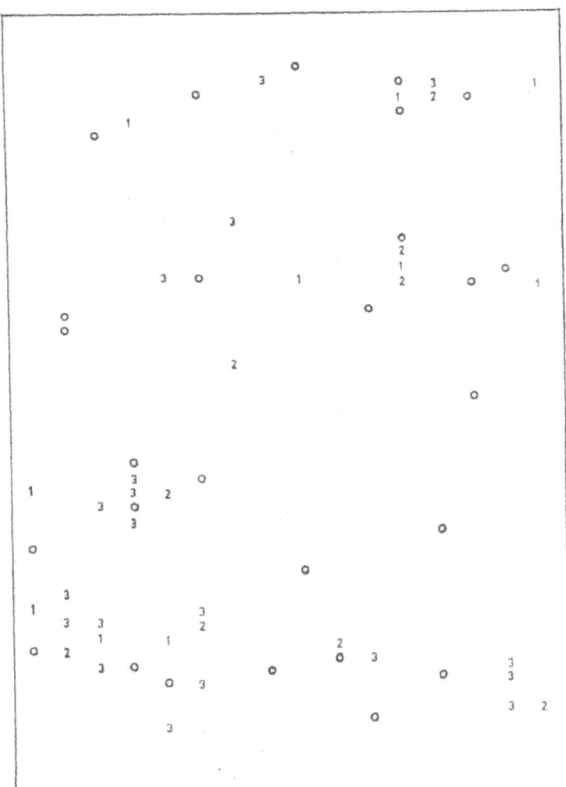

Figure 24. The map shows the spread of VNPV in a vanilla plantation from 0 (first month) to 3 (4th month).

Several plots of vanilla were mapped and the number of new infected plants were recorded every month. Figure 24 shows the spread of the VNPV in the plantation over a 4 month period. VNPV spread rapidly along the rows as well as between the rows of vanilla. When tested, the spread was found to be influenced by the proximity of infected plants. Therefore it is not random but contagious (P = 0.025).

39

The distribution of infected plants were not random but clumped, clustered or contagious. The "clumping" effect proves that the presence of infected plants increase the probability of the neighbouring plant being infected. Tests were significant at P=0.025. Two other plantations were also monitored during the same period. VNPV infected plants were also found to be "clumped", thus suggesting that infected plants increase the chance of neighbouring healthy plants being infected as well. Tests were significant at P=0.05. Vanilla growers normally work along the rows of vanilla and it is highly likely that they, unwittingly, spread the VNPV along the rows of vanilla through sap transmission on hands or tools (P=0.025). Aphids were also observed to probe on the vanilla leaves in the laboratory, so it is highly likely that they also spread the VNPV between the rows of vanilla (P=0.025).

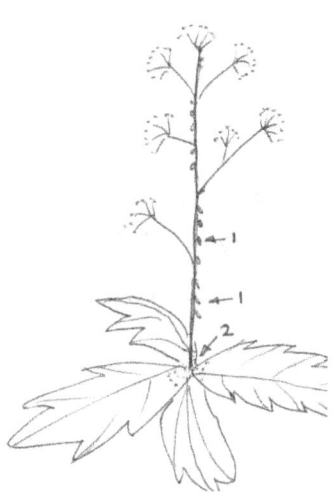

Figure 25. A large number of the weed *Emilia sonchifolia* were found in the infected plantation, with large resident aphid (1 & 2) populations. Growers were advised to remove all aphid hosts from within their vanilla plantations.

New vanilla plantations were inspected for the presence of the VNPV symptoms. All infected planting material also produced symptomatic shoots. It proved conclusively that long distance spread of the VNPV is mostly infected cuttings. Large numbers of cuttings are often shipped from Vava'u to Tongatapu and probably included VNPV infected cuttings. A full discussion of the effect of infected cuttings in long term transport of VNPV is in **"Epidemiology of Vanilla Necrosis Potyvirus in *Vanilla fragrans* (Salisb) Ames plantations in the Kingdom of Tonga"** by Semisi Pone.

LINEAR DISTRIBUTION OF VNPV IN VANILLA ROWS.

The spread of VNPV along the vanilla rows was assessed using the doublet analysis (van der Plank, 1960) where n plants are investigated in a sequence, and of these u are diseased, then the expected number of doublets (d) is given by $d = u(u-1)/n$.

A similar equation was proposed by Madden and Campbell (1986) as $E(D) = m(m-1)/N$ where $E(D)$ is the expected number of doublets, m is the number of infected plants in a row, and N is the total number of plants in that row.

For rows with $N = 20$ a test of randomness is given by

$Z = (D-E(D))/s(D)$ where $s(D)$ is $\{[m(m-1)/N](1-2)/N)\}^{1/2}$.

If $Z > 1.64$, the null hypothesis of randomness is rejected in favour of clustering $(P = 0.05)$.

SPATIAL DISTRIBUTION OF VNPV.

The spatial pattern (spread in 2 dimensions) of infected plants will be clustered or clumped if the infection of a given plant increases the probability of another plant nearby being infected. Vanilla plantations were divided into quadrats and the variance (s^2) and the arithmetic mean (u) of the infected plants in each quadrat are compared. When $s^2 < u$, the distribution is that of a positive binomial or regular. When $s^2 = u$, it is a poisson series or random. When $s^2 > u$, the distribution is a negative binomial or contagious (Elliot, 1977).

A test for significance is $x^2 = s^2 (n-1)/u$. When x^2 is greater than the corresponding value in a chi-square table with n-1 degrees of freedom and significance level P, the poisson distribution is rejected in favour of clustering (Madden and Campbell, 1986). Contagious or clustered distribution means that there is an agent like an aphid or the growers themselves, spreading the virus between vanilla plants.

HEALTHY PLANTING MATERIAL.

The importance of selecting healthy planting material should be emphasised. In a survey of new plantations, less than 2 years old, it was proved that ALL symptomatic new shoots develop from symptomatic cuttings. None of the healthy looking cuttings developed any shoots with disease symptoms.

LIMITATIONS TO VANILLA PRODUCTION IN TONGA.

1. CLEAN PLANTING MATERIAL

Selecting clean vines to start a vanilla plantation is the key to success. Figure 26 shows a healthy vigorous plantation with very good management. Dry mulch is used in this plantation. A large number of vanilla plantations especially on Vava'u use live grass as mulch. Live mulch also works very well but the grass,

a species known locally as "vailima", can be overgrown sometimes and causes difficulty for workers going through the plantation.

Figure 26. A healthy vanilla plantation with grass mulching between plants. Vaini Research Station, MAFF, Tonga.

Clean planting material is one of the control methods recommended especially to new vanilla growers. In older plantations, symptomatic plants are removed and burned or buried. Healthy cuttings are used to replace them. Growers are advised to make sure cuttings are

43

taken from healthy non-symptomatic plantations only. Vanilla plants with mottled leaves should never be used as planting material for a new plantation. Vanilla plants exhibiting mottled or yellowing leaves are often thinner and less vigorous than green healthy plants.

2. MULCHING

Vanilla roots are mostly found in the first 1-2 inches of the topsoil and a cover of dry plant material or "live grass" contributes to keeping the soil moist.

It has also been proven that mulching the vanilla plants produce more vigorous plants that bear more beans per vanilla plant. Dry mulch or live grass mulch are both recommended. The advantage of dry mulch is that you don't have to wade through a "grass jungle" to do your plantation work as the grass can sometimes be up to 2 feet high.

3. LABOUR

Vanilla growing is a labour intensive industry. However, it is very profitable. It has been reported in the newspapers (Tongan Times) that a 40 acre plantation in Vava'u produced $TOP 900,000 worth of green vanilla beans in one very productive year.

The most intensive vanilla practice is artificial pollination. Insects do not pollinate vanilla flowers outside its country of origin (Mexico), where it is pollinated by a local bee. They have to be pollinated by hand. Addition of mulch, about once a year, is another. Flower initiation and training of vines is a distant third. Curing is also a very long process which takes up to 3 months but growers can opt to sell their mature beans instead of curing it themselves.

4. DISEASE CONTROL

There are no major disease problems of vanilla in Tonga apart from VNPV. It is the only single most destructive disease of vanilla plants in the Pacific Islands and, I suspect, Indonesia and other vanilla producing countries. An ELISA test with VNPV antiserum can easily prove the presence of the virus. A potyvirus group test is also available from Sigma, USA.

Electron microscopy, at the University of Auckland, found CyMV and ORSV particles. Filamentous, flexuous virus particles of around 799 nanometres in length were also found in symptomatic leaves. That usually means it is a virus in the Potyvirus Group (Pearson, MN and Pone, SP, 1988). Further research in Tonga and the University of Auckland proved beyond reasonable doubt that VNPV is the virus causing the problems.

A control strategy was developed with Dr MNPearson of the University, myself and MAFF, Tonga extension/advisory staff. All symptomatic plants in the Kingdom were pulled out and destroyed. Clean vines were used to plant new ones. Cultural practices in the plantations were designed to avoid spreading the virus (Pone, SP and Pearson, MN, 1989).
It was interesting to note that only 3 years later the exported vanilla beans jumped from 20 tonnes to 70 tonnes. I cannot say with absolute certainty whether this was due to the nationwide VNPV disease control strategy implemented or the increased acreage due to the MAFF Vanilla Development Project. Perhaps a combination of both.

5. HARVESTING.

Vanilla beans are harvested when they turn light green to yellow at the distal end. This indicates that it is mature for curing. It has been noted during the virus research programme that VNPV infected plants which produce beans usually lose the beans, they ripen prematurely or turn necrotic. The distal end turns black instead of greenish/yellow. I have cured those beans and send them for testing at overseas laboratories

(France). Vanillin content is usually low (less than 2%), although they do smell like normal cured beans. In normal beans vanillin content can be between 2+%-4%. Diseased beans tend to weigh less, appear brown and "woody" and less shiny.

6. CURING BEANS.

Curing vanilla beans is a very long process, but I have found it a very enjoyable and sweet smelling exercise. Diseased vanilla beans (Figure 27) do not cure as well as the normal ones (left) . They dry out faster, have lesser weight and lesser vanillin content.

Figure 27. Vanilla beans which started rotting at the distal end due to VNPV (left) and A quality vanilla beans (right) after curing. They look very similar except A quality vanilla beans are darker, heavier, shiny with crystals on the surface and higher vanillin content. Beans affected by VNPV are "woody", lighter in weight and colour with low vanillin content (generally less than 2%).

VNPV affected vanilla beans will fetch a low price in the market, but they are still usable. The A grade beans appear dark, shiny, with crystals of vanillin on the surface and with high vanillin content of between 2.5-4%.

However, my recommendation was to destroy ALL vanilla plants that show any symptoms of vine or leaf necrosis. That is the only way to get rid of the VNPV infected plants, AND use "healthy" looking vines with no mottling, chlorosis or necrotic spots as new planting material.

7. STORAGE

Madagascar, the largest natural vanilla producer in the world, usually have a stock of more than a 1,000 tonnes of cured beans at any time. That is during my years of work (1985-1992). They control the world price by how much they sell in the world market in any one year.

Vanilla is one of the few high value natural products that can be stored for a long time. It is something that Pacific Islands should think about. They have small land areas, perfect weather and abundant labour. Vanilla can be produced and stored, to be released to the market when the price is high.

8. EXPORT MARKETS

Natural vanilla is preferred over synthetic "vannaline" by the producers of high quality flavourings, restaurants and high flyer markets. They prefer to pay high prices for the best natural vanillin to flavour their products.

Synthetic vanalline is very cheap and more widely used for the lower end of the confectionery market. The "high flyer" markets or the "posh" restaurants have appearances and their taste buds to satisfy and should keep the natural vanilla suppliers in business for a long time at higher prices.

Note: A new vanilla plantation will take 2-3 years to start producing beans. Well managed plantations will keep on producing for up to 12 years or more. A 4 acre vanilla plot can earn up to $TOP 20,000 or more per year depending on management and weather conditions.

CHAPTER 3. KAVA CUCUMBER MOSAIC CUCUMOVIRUS.

Kava (*Piper methysticum*) is a popular traditional social drink in many Pacific Island countries including Tonga, Fiji, Samoa, Vanuatu, Hawaii and some Micronesian countries. It was used mostly for ceremonial purposes but is becoming a rapid replacement for alcohol in many metropolitan Pacific communities including New Zealand, Australia, Hawaii and mainland USA.

Figure 28. Kava plantlets being prepared in the "screenhouse" for replanting to the field. Vaini Research Station, MAFF, TONGA.

The discovery of "Kava dieback" many years ago, probably 30 years or more, let to a lot of research around the Pacific to try and identify the causal organism. Vigorous Kava plants usually "dieback" from the shoot hence the name. The shoots become chlorotic with sunken spots on the leaves which turn necrotic. The necrosis spreading rapidly on the leaves and down the stem. New shoots collapse and regrow but eventually succumb to the disease.

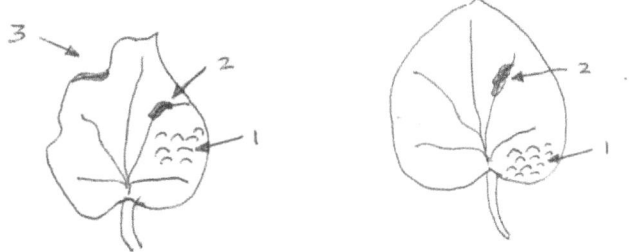

Figure 29. Newly infected kava plants shows leaves with "crinkling" (1), veinal chlorosis (2) which is usually yellow/orange in colour and sometimes puckering of the leaf edges (3). Some leaves also show faint mottling on the lamina.

A joint project by MAFF, Tonga and ACIAR (Australian Centre for International Agriculture Research) finally solved the problem after decades of unsuccessful attempts at identifying the disease by many of the regions best Scientists.

Chlorotic sunken spots on young Kava leaves let to my hypothesis that it may be a virus. Leaves were sent to Australia where electron microscopy examination of the Kava sap found particles similar to those of Cucumber Mosaic Cucumovirus (CMV) in the Kava sap. It was the first time CMV has been reported from the Piperaceae.

The whole direction of the project, at the time, changed from fungal to viral research. Dr Richard Davis and Professor John Brown of the Australian National University led the Australian team, co-ordinated by Dr Paul Ferrar of ACIAR. The MAFF Research Division, with funding from the Tonga-German Plant Protection Project, provided the counterpart work which was mainly Enzyme-Linked Immuno-Sorbent Assay (ELISA) tests of kava plants to confirm the presence of CMV. ELISA Tests were optimised with

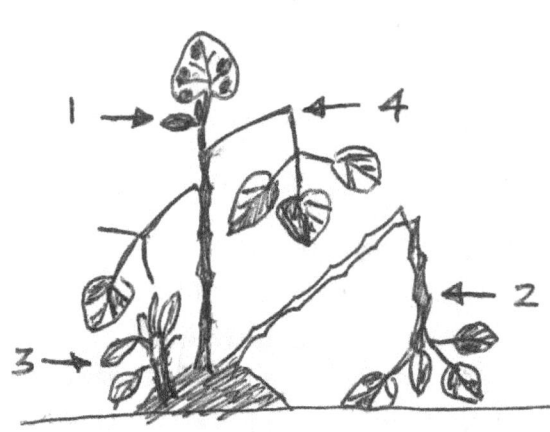

Figure 30. A kava plant in the advanced stage of dieback. Most leaves and stem are necrotic (1 & 2), new shoots are stunted (3) and petioles break (4). Note broken branch.

antibodies to CMV purchased from Sigma, USA. I had commented to Dr Richard Davis, during our frequent ELISA test exercises, that the reaction of the CMV in kava is not as strong as the reaction of VNPV in vanilla to the ELISA tests. It could be a new CMV strain or probably the difference in the enzymes used in the ELISA tests. We used alkaline phosphatase in the case of VNPV and horseradish peroxidase in the case of CMV. The results of the research was published by Davis R.I., Brown J. F., and Pone, S.P. (1996).

KAVA CUCUMBER MOSAIC VIRUS CONTROL.

The best way to avoid CMV infected plants is to select healthy and vigorous looking plants and use them as planting material. Do not use any plants that have yellow spots on the leaves or show any symptoms of dieback. CMV is systemic inside the plant and even though shoots may look healthy, CMV virus particles may be present in the young leaf tissue. Once the plantlets start growing virus particles rapidly multiply and ultimately destroy the new shoots.
The healthy shoots can be grown in plastic bags or pots before planting into the field. Cut the shoots into one node size and seal the ends with wax or fungicides to avoid rotting and plant in the pot or plastic bag. Use "light soil" or mix 50/50 with sand. Make sure pots or plastic bags have "water holes" to allow free drainage or the pot/plastic bag may become waterlogged. Transplant plantlets into the field when they are about 15 centimetres high. Make sure the field is well prepared. Ploughing twice and furrowing will soften the soil and allow the kava to grow quickly.

As a virus control method, remove all herbaceous weeds from the immediate area and inside the plantation, at all times, as they will harbour aphid populations, which may transmit CMV between the plants, if it gets introduced into the plantation. There is no chemical control for CMV at the moment.

KAVA TRADITIONS AND SALES...

Kava is widely consumed in Tonga and in Tongan communities overseas. It is a very important traditional crop. Most Tongan ceremonies from the King's kava circle…called the "Taumafa Kava"…where Noble and Chiefly titles are confirmed…. to village weddings use Kava as the only ceremonial drink. Kava is also consumed in many Pacific countries. In Vanuatu, Kava Clubs are known as "Nakamal" and very strong Kava is consumed in them. Visitors are advised not to drink more than 3 coconut shell cups of Kava.

Figure 31. Kava plantlets being prepared in the glasshouse for planting to the field or sold to kava growers as planting material. MAFF Research, Tonga.

Fiji, Samoa, Hawaii and some Micronesian countries also drink Kava. It is becoming a very important commercial crop in many Pacific Islands.

Its commercial value is the main reason why the Tongan, Fiji and Samoan Governments were very keen to find a control method or control strategies for Kava Dieback, because a large number of people and the Government make a lot of money from its production and sale.

LIMITATIONS OF KAVA GROWING IN TONGA.

1. PLANTING MATERIAL AVAILABILITY

Using healthy, virus free, planting material is the best way to avoid Cucumber Mosaic Cucumovirus (CMV). If there is no source for the virus, growers can have a relatively healthy plantation.

2. DISEASE CONTROL

CMV is aphid transmitted. That is why it is important to eliminate all sources of the disease near the plantation because aphids can transmit the virus from plant to plant by landing and probing on the leaves. Most viruses are transmitted by aphids which carry the virus particles on their feeding stylet. Cultural practices inside the plantation may not transmit the virus as in the case of VNPV because there is no sap transmission between plants.

3. LABOUR COSTS

Kava growing does not require a lot of labour apart from planting, weeding and harvesting. Kava plants are normally planted with other crops and does not need any further inputs until harvest. No fertiliser or chemicals are required. Like vanilla, kava is a totally organic product.

4. MARKETING

Marketing Kava overseas is not necessary as almost all of the local production are consumed by Tongans in Tonga and overseas. Kava clubs consume the most followed by kava ceremonies like weddings, funerals and so on.

Some Kava are also sold to pharmaceutical companies for making drugs to calm the nerves.

CHAPTER 4. ZUCCHINI YELLOW MOSAIC VIRUS (ZYMV) AND SQUASH EXPORT FROM TONGA.

Squash export to Japan was one of Tonga's largest industry by volume and earnings in the late 1980s and early 1990s. It reached a peak of more than 20,000

tonnes per year in the early 1990s. In 1991, ZYMV was discovered on squash plants in some plantations. It rapidly spread and totally destroyed some plantings. All the young squash leaves turned yellow and their growth stunted by the virus. In some

Figure 32. Early ZYMV infection of squash planlets shows faint veinal chlorosis (plant on left). Growers were advised to replace them with healthy plantlets from other mounds. Note aluminium foil as a "reflector" to repel aphids from landing and probing on the squash plants. MAFF Research, Tonga.

cases, it was 100% loss to the growers. Some research were carried out at MAFF Research, Tonga and control strategies were put in place to try and reduce the damage by ZYMV.

Research results found aphid numbers to be significantly reduced by "reflective mulch". Aluminium foil, white plastic and dry grass were tested. All were effective in reducing aphid numbers around the young squash plants. Dry grass was recommended to growers as they are readily available and cost nothing. Reflector mulch was advised to growers to repel aphids from landing and probing on the squash plants and transferring the ZYMV disease/virus particles. Aphids pickup the ZYMV virus particles on their stylet from nearby infected plants and infect new healthy plants with it by probing/feeding on their leaves.

ZYMV control strategies were aimed at delaying ZYMV infection of squash plants until fruits are formed. Observations suggest that even when the vine is infected later, marketable fruit is still obtained.

All the reflective materials tested were found to be significantly better than the bare ground in reducing the number of aphids in the immediate vicinity. Reflective mulch is being used in many countries for that purpose. It has been suggested that the reflected rays of the sun, off the mulch, confuses or "repels" the aphids.

Barriers were also suggested to growers. Corn was the most popular choice as growers can harvest the corn and sell them as well. Rows of corn were planted at about 10 metre spaces in the squash plantation, perpendicular to the prevailing wind direction to "trap" aphids as they move into the squash plantation. Aphid movement is usually affected by wind direction. When the aphids probe on the corn leaves they "lose" their virus load in that process. Aphids can also land and colonise the corn instead of the squash. All growers who planted corn barriers were pleased with the results of the squash harvest and the additional cash from the corn sales.

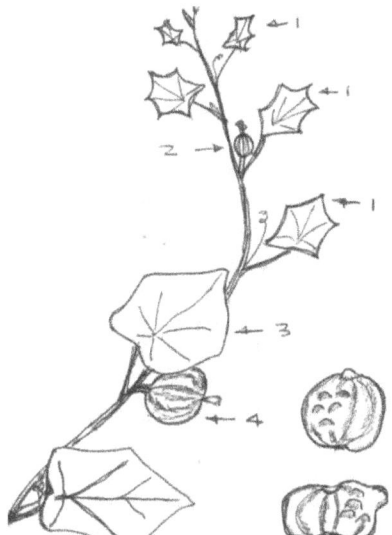

Infected squash vines do not produce any marketable fruit. They are usually deformed, undersized, with lumps on the surface. However, the older part of the same vine, with no symptoms on the leaves, usually produce large marketable fruit. It is an important observation as some of the control measures

Figure 33. A ZYMV infected squash vine showing the deformed, chlorotic young leaves (1), new fruit (2), older non-symptomatic leaf (3) and fruit (4). Note deformed fruits on the side. Lumps are shown on the surface.

59

were aimed at delaying ZYMV infection until the fruits form. Growers were advised to cut off the infected end of the vines with symptoms.

Cleaning infected alternative hosts from around the new plantation was emphasised to growers as old cucurbit plots like watermelons, cucumber and zucchini sometimes act as a source of inoculum for new squash plantations. Many squash plots were destroyed by ZYMV before fruiting due to the presence of old infected cucurbit plantations nearby. Aphid populations are normally high within and around squash plantations due to the presence of herbaceous weeds. There are 8 recorded species of aphids in Tonga. The most prevalent being *Aphis gossypii*.

There were hundreds of squash growers so it was a major operation via radio programmes, advisory and farm trainings and extension work. Despite the losses, growers quickly got on to the practice of avoiding the virus by removing all virus

Figure 34. A neat squash plantation in Port Vila, Vanuatu. Squash was a popular export crop for many Pacific Islands.

sources around their plantations which became very

effective in reducing and controlling virus spread and damage on a national scale. Yield and total production went up instead of down. It was one of the most successful operations by MAFF Research in terms of disease control as squash were planted and harvested within 5 months so there was little time to work with as in the case of bananas, vanilla and kava.

ZYMV is no longer a major problem in Tonga since the end of the squash export operation. As with most diseases, it grows exponentially as a response to the increasing acreage. When the acreage is reduced the disease crashes. That is why cucurbits can still be grown successfully in Tonga and around the Pacific, now and in the future. This includes watermelons, zucchini, cucumber, rock melon and others. Current cucurbit acreage is low so disease spread and effect on cucurbit plantations is minimal.

Figure 35. Harvests from a trial plot of "Kurijiman" squash variety grown at Apia, Upolu, Samoa. It was an attempt to test this variety in Samoa for export to Japan. High virus disease pressure caused early infection of the two trial plots.

Figure 35 shows the results of a private "trial" to test marketability of squash from Samoa. Although the Japanese buyers were keen to sell the squash from Samoa, the project was cancelled due to "high disease" pressure. Squash plants were infected by the ZYMV before they flower. This means that fruits will be deformed if they form at all. Normal fruits were still obtained from some plants, but the risk of heavy losses to the Samoan squash growers was too high. So the project was cancelled.

THE WORLD DISTRIBUTION AND CONTROL OF ZYMV...

ZYMV is a very destructive disease of cucurbits. It is present in most countries of the world that grows cucurbits. In the Pacific Islands it is the most limiting factor in cucurbit production. That includes watermelons, cucumber, zucchini, rock melon and others. The best way to control it is to;

1. **Remove all alternative hosts from the immediate area you intend to plant in.**
2. **Use reflective mulch like dry grass.**
3. **Remove herbaceous weeds that harbour aphids from your planting area.**
4. **Remove infected plants within your plantation as soon as they show symptoms on the leaves.**
5. **After harvest destroy all hosts to reduce source of infection.**

CHAPTER 5. ANTHRACNOSE OF YAMS AND TREECROPS.

Anthracnose is one of the most widespread diseases in Tonga. It can be found in almost all crops and trees. It attacks crops and sometimes as a secondary pathogen. It causes flower damage to most tree crops including mangoes, avocado, Polynesian lychees, bananas and many other fruit trees.

It is caused, mainly, by the fungus *Colletotrichum gloeosporioides* and related *Colletotrichum* species.

Information on anthracnose infections and control in this chapter are derived mostly from my own experiences otherwise references are quoted.

YAMS

One of the most important crops attacked by anthracnose is yams. The early yams in Tonga, which are planted between May and October are usually affected namely the most popular variety "kahokaho". They have long tubers, up to 5 feet or longer. Chemical control was used. Benlate 50%WP and Manzate are sometimes used in combination when black spots (known as **mahunu**) start to appear on the yam leaves.

Spraying continues, every two weeks, until the plants are strong enough to withstand the disease, which is about 2-3 months. The spray programme is then discontinued.

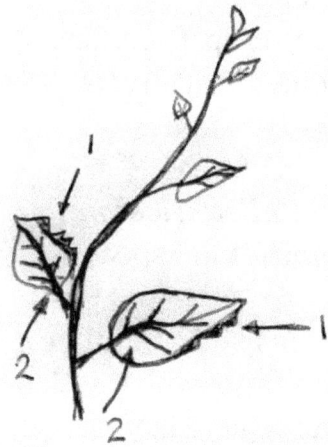

Sometimes "spot" sprays maybe applied during wet weather to reduce damage to yam leaves. Some other early yam varieties like **"kaumeile"** may be affected but not as bad as kahokaho.

The Tongan "lesser yams" known as **"tokamui"**, which

Figure 36. Symptoms of anthracnose on "kahokaho" leaves. Note necrotic leaf edges (1). Usually where water collects during rain showers. Veins are also pronounced and turning necrotic (2). The popular variety "kahokaho" is one of the most susceptible varieties.

are planted between October and December are not normally attacked by anthracnose. They have round and short tubers of less than 2 feet in length.

TREECROPS

Anthracnose also attacks mango flowers at about the same time it affects early yams. The mangoes normally starts flowering around July to August. If it is a dry year there will be a "bumper" crop. If it is wet, the loss is usually 80-100%. There will be hardly any mangoes for Christmas. They usually fail to set and fall off without developing any further.

Tava (***Pometia pinnata***), also known as Polynesian lychees, flowers appears to be affected in the same way as mangoes, by anthracnose. Sometimes Tava trees, in Tonga, fail to fruit for many years. Fungal hyphae growth can sometimes be seen on the blackening flowers as the disease destroys it.

Figure 37. Mango flowers affected by anthracnose shows "blackening" of some parts (1) of the panicle. The stalk also turns necrotic (2).

Banana is also highly susceptible to anthracnose. Although only associated with over-ripe bananas, anthracnose can cause "premature" ripening under certain conditions.

Anthracnose was also mentioned as one of the post-harvest problems of bananas exported from the Pacific Islands to the New Zealand market.

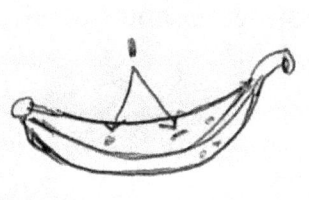

Figure 38. Anthracnose "spots" on bananas (1).

Mango is a very

popular fruit worldwide, but it is affected by anthracnose in most countries. The anthracnose attacks both the flower panicles and young and ripening fruits.

Figure 39. Anthracnose "spots" on mangoes (1), left.

Anthracnose necrotic spots on fruits (Figure 38 and 39), especially bananas and mango are usually signs they are "overripe" but may also contribute to marketing problems overseas, if fruits develop these spots at an early stage. Avocado and many other fruits are also affected by anthracnose in many countries.

CHAPTER 6. TISSUE CULTURE AND HURRICANE RECOVERY IN THE PACIFIC.

The Tissue Culture Laboratories at the University of the South Pacific in Samoa and the South Pacific Commission (now called the Secretariat for the South Pacific Community) in Suva provides excellent services to the Pacific region by preserving all the popular crops in tissue culture. These germplasm collections are being used to supply Pacific Island countries during times of need, like hurricane recovery and research.

In some cases susceptible but popular rootcrop varieties, for example the sweet potato variety **"Tongamai"** which was very popular in Tonga but was highly susceptible to **"sweet potato scab"**, can be stored in tissue culture for many years until scientists can work out how to control the disease that affects them. The Tissue Culture Laboratory at USP was under a European Union funded project called the Pacific Regional Agriculture Programme - Project 7(PRAP 7). It collected yams, cassava, bananas, giant taro, taro, dasheen, vanilla, sweet potatoes and many other species and stored them in tissue culture. They were supplied to Pacific Island countries after hurricanes for agriculture recovery projects and for research.

One of the obvious advantages of Tissue Culture is its "disease free" status. They can be "pathogen tested" before being stored in Tissue Culture almost indefinitely for the Pacific Islands. Countries have

Figure 40. Tissue cultured banana plantlets from the laboratory, ready for planting to the field. Cavendish and other banana varieties were supplied by ACIAR, Australia to MAFF Research, Tonga.

access to the germplasm collection at any time. It is the best way of distributing planting material between countries in the Pacific Region.

Large recovery programmes such as the efforts to revive the Taro Industry in Samoa is possible with Tissue Culture. Millions of plantlets can be produced every year and distributed to the growers. Disease control and industry recovery is possible within a few years from the first epidemic. Co-operation between all parties is the key to its success, if resistant varieties of taro can be obtained. This may require political agreements between Pacific Island countries to share their taro germplasm.

Figure 41. A screen house at the Banana Research Institute, Taiwan, ROC. Tissue cultured bananas are produced every year for the Taiwanese Banana Industry to reduce disease prevalence in the field by providing clean planting material.

The two regional laboratories may have as many as 200+ varieties of taro (colocasia), 200+ varieties of sweet potato, 10+ varieties of bananas, 5+ varieties of yams, 3+ varieties of cassava and many other crops varieties. Obviously if all

Figure 42. Tissue culture bottles being used to store germplasm of popular Pacific crops at the University of the South Pacific, Alafua Campus. The project was funded by the European Union through its PRAP initiative. IRETA implemented the project.

69

the popular crops of the Pacific Islands are stored in a germplasm collection, it will be very useful, not only for hurricane recovery and research but also for future generations. The Taro Leaf Blight destruction of taro in Samoa is an obvious example of why there is a need for a germplasm collection in the Pacific. Resistant taro varieties can be stored and supplied to growers when needed.

There are many resistant varieties to TLB in Papua New Guinea, Solomons and some other islands but they have not been stored in culture, yet.
If the resistant varieties were present in the Tissue

Culture labs at USP and SPC they would have been made available to Samoa and their Taro Industry would have recovered completely. However, twenty years later it has clearly not. Despite talk during the efforts to assist Samoa control the TLB problem during 1993-1996, no one has been able to help them with the Taro Industry recovery. It may

Figure 43. Grande Naine cavendish banana variety from ACIAR, Australia being tested at the MAFF Research Station in Tonga.

never recover as Fiji has total control of the taro market at the moment. Banana trials in Tonga shows the best use of tissue culture in the region. It was possible to

revive the Samoan Taro Industry in the same way. The tissue culture lab at the Coffee Research Station in the Papua New Guinea highlands was one of the laboratories considered for the **"Taro Tissue Culture Project"**

Figure 44. Tissue cultured vanilla plantlets produced in tissue culture after being ELISA tested for virus presence. It was one of the strategies to control VNPV in Tongan vanilla.

which was aimed at introducing "**disease free**" resistant taro varieties to solve the taro leaf blight

problem in Samoa. Although ACIAR had approved $AUD 1 million for the project, it was cancelled when I left the South Pacific Commission. No one was able to implement the project. The next Plant Protection Advisor/Co-ordinator for the SPC-PPS did not assume his post until some months later.

Figure 45. The Tissue Culture Laboratory at the Coffee Research Institute in Papua New Guinea is one of the best in the Pacific Region.

A lack of foresight by everyone concerned. Samoa has lost up to $ST300 million in taro exports since 1993.

There are many world class facilities in the Pacific Region for Tissue Culture work and germplasm collections. They just need to co-ordinate their work to avoid duplication and save funds for other work. As mentioned, a political agreement should also be made before germplasm is shared between countries as many countries are very "protective" of their "super species", like TLB resistant, high yielding taro in Papua New Guinea, for example.

Note: During the writing of this book, I came across some information in the Food and Agriculture Organisation of the United Nations (FAO) website which mentions that "resistant taro varieties" to Taro Leaf Blight are being tested in West Africa after an epidemic over there. The resistant varieties came from the Tissue Culture Laboratory at the Secretariat for the Pacific Community (SPC) in Suva. However, there is no mention of any work done to test them in Samoa. The FAO regional office in Apia, Samoa does not appear to have any information on TLB resistant work in Samoa, in its website.

CHAPTER 7. BIOSECURITY AND NATIONAL PLANT PROTECTION ORGANISATIONS.

The introduction of many destructive diseases like taro leaf blight, banana black leaf streak, zucchini yellow mosaic virus and tristeza virus to the Pacific has clearly demonstrated the problem of biosecurity in the region.

The establishment of the Pacific Plant Protection Organisation (PPPO) in 1994 by a Resolution of the 34th South Pacific Conference was an attempt to improve biosecurity in SPC member countries. The Legal Division of the Food and Agriculture Organisation of the United Nations helped establish the PPPO by writing the resolution together with the staff of the SPC Plant Protection Service and assistance from the New Zealand Department of Foreign Affairs. The FAO legal counsel had worked together with the SPC Plant Protection Advisor and SPC member countries to draft the resolution which was accepted and passed by the Conference, after 8 years of discussions!. The PPPO became a member of the International Plant Protection Commission's (IPPC) Regional Plant Protection Organisation (RPPO) initiative. RPPOs met in Rome at the Food and

Agriculture Organisation (United Nations) headquarter every 2 years, for technical consultations.

In addition, a panel of experts on Biosecurity also meet every other year when the RPPOs are not meeting.

It was a chance for a representative from the 22 small island nations in the Pacific to meet and talk with representatives from other RPPOs from other regions around the

Figure 46. Participants of the 7th Technical Consultation among RPPOs, SPC Headquarters, Noumea, New Caledonia. Dr Niek van der Graaf, Chief of Plant Protection, FAO; Ati George Sokomanu, Secretary General, South Pacific Commission and Dr John Hedley, IPPC are shown on the front row. Author is on left of picture.

world. For example, The European Plant Protection Organisation, the North American Plant Protection Organisation, the Asia-Pacific Plant Protection Commission and others from South America, Africa and the Caribbean.

One of the advantages of RPPO Technical Consultations was the obvious networking and sharing of information. For example, a disease like Taro Leaf Blight may cause an epidemic in the Caribbean Islands, if introduced there. They can report this to us at the PPPO within minutes of its discovery. Biosecurity measures are quickly put in place at Ports and Airports, in the Pacific Islands to prevent its introduction. Taro is a very large industry and stable food in the Pacific Islands and has to be protected from foreign diseases.

The South Pacific Commission Plant Protection Services provides the Secretariat for the PPPO at no cost to member countries. The Plant Protection Advisor/Co-ordinator of the SPC-PPS acts as the Chief Executive and represents the PPPO at all international meetings.

Figure 47. Signing of the MOU between the PPPO and APPPC. Mr Poloma Komiti (left), SPC Director of Programmes and Professor Shen (right), Chief Executive of APPPC, signed the MOU. Mr Bob Ikin of Australia, first Chairman of the PPPO meeting watches (middle). Author stands from the back.

The first meeting of the Pacific Plant Protection Organisation was held at the Tanoa Hotel in Nadi, Fiji in February,

1996. Its work programme for the next 2 years was discussed and approved by member countries.

Meetings were held "back to back" with the Regional Technical Meeting on Plant Protection (RTMPP)to lower costs. The FAO Database on Plant Diseases and plans for a Global Network were also discussed. Participants were shown how the network would operate by FAO Epidemiologist, Dr C A J Putter and others. The

Figure 48. Shaking hands after signing the MOU between the Pacific Plant Protection Organisation and the Asia-Pacific Plant Protection Commission at the Tanoa Hotel, Nadi, Fiji.

SPC Plant Protection Database was to be linked to

FAO Headquarters in a new TCP (technical consultation project). A memorandum of Understanding (MOU) was signed by the Pacific Plant

Figure 49. Former SPC Plant Protection Officer, Mr Bob Macfarlane, training NPPO staff in French Polynesia on how to use the SPC-PPS database.

Protection Organisation

(PPPO) and the Asia-Pacific Plant Protection Commission (APPPC) to work closer together on Plant Protection issues in the region. The APPPC Chief Executive and the SPC Director of Programmes got approval from the Director General of FAO and Secretary General of SPC, respectively, to sign the MOU on their behalf.

The APPC is the largest RPPO in terms of number of members, population and area. All countries of Asia and some countries from the Pacific are members of the APPPC. The PPPO, by comparison, only has 27 members at the time. These include American Samoa, Australia, Cook Islands, Federated States of Micronesia, Fiji, France, French Polynesia, Guam, Kiribati, Nauru, New Caledonia, New Zealand, Niue, Northern Mariana Islands, Papua New Guinea, Pitcairn Island, Republic of Marshall Islands, Republic of Palau, Samoa, Solomon Islands, Tokelau, Tonga, Tuvalu, United Kingdom, United States of America, Vanuatu, Wallis and Futuna.

Note: It would be in the best interest of the countries of the Pacific to have the "Pacific Curtain" established in some form. The advent of the email might have rendered it redundant, but some kind of agreement or general awareness of it would be good for border control, disease and pest prevention and control.

It was part of the overall strategy establish the PPPO

connections around the Pacific Islands. The first PPPO meeting was also a chance to show the participants how the Global Database Network will operate from FAO Headquarters. The SPC-PPS Database, together with free

Figure 50. Participants of the first Pacific Plant Protection Organisation meeting, Tanoa Hotel, Nadi, Fiji learning how to use the Plant Protection Database from FAO/SPC. Atoloto Malau of Wallis and Futuna is in the foreground. Konrad Engleberger from Tonga MAFF/ Quarantine is also shown on picture.

computers, was provided by the SPC Plant Protection Service through the SPC/EU Pacific Plant Protection Project to 8 ACP countries and 3 French Territories. The other countries will also receive computers and the database later. The European Union only funded countries included in the Lome Convention and European Territories. Counterpart funding were sought from New Zealand and other countries to include all 22 Pacific Island countries and territories in the network, if possible. I left SPC before the project was completed. It was also one of the important projects that was probably abandoned due to the change in staff leadership of the SPC-PPS.

I cannot praise the work of Regional Plant Protection Organisations enough. The losses due to Taro Leaf Blight and Banana Leaf Streak demonstrates the value of being vigilant and keeping plant diseases and insects out of the region. There are still diseases like viruses on pawpaw and cassava* and so on which are not in the Pacific Islands. If they are introduced they will wipe out these very important staple food plants.

One example is citrus in Tonga. Tristeza Virus has wiped out more than 90% of mandarins and oranges in Tonga. It was only introduced in the middle to late 1980s. Ten years later most of the citrus on Tongatapu have disappeared. There is still constant regrowth from the seed bank in the ground but soon these will be exhausted.

---

*There is a report from Sri Lanka that Cassava Mosaic Virus is becoming a problem there, as I write this book. It may reach epidemic proportions if not controlled.

Note: Banana leaf roll is present only in some countries of the Pacific including Papua New Guinea and probably part of Solomon Islands. It may have been introduced from Asia via Indonesia or Irian Jaya. Papua New Guinea may have to be extra vigilant on its border with Indonesia as there are many pests in Indonesia not present in PNG.

Figure 51. One of the serious banana pests to be avoided is banana leaf roll shown here on plantains in Papua New Guinea.

There are also pests which are present in some parts of the Pacific, but not in others, like the banana leaf roller, for example. It is present in Papua New Guinea but not in other Pacific countries. It does a lot of damage to banana and plantain leaves and therefore should be quarantined on a regional basis. Figure 51 shows what kind of damage the banana leaf roller can do to bananas and plantain. Total loss of foliage may occur in some cases.

One of the important discussions at the RPPO meetings in Rome was to establish a global database of plant diseases and insects with an internet link to all RPPOs and the IPPC at FAO headquarters, the training at the first PPPO meeting being one of the various actions taken to establish it. All the RPPO contacts will learn of plant disease and insect movements on a global scale and act accordingly. It was the early days of the internet and everyone was keen.

I had hoped we will have such a tool in the Pacific which we can combine with other activities to form what I call the "Pacific Curtain", to keep pests out of the region.

The PPPO should be represented at all RPPO meetings, including the Technical Consultation in Rome. Networking and establishing contacts are very important for the isolated islands of the Pacific. I had attended the 18th and 19th APPPC meetings and also visited some of the agricultural areas in China and the Philippines, respectively. I got expert information on agriculture in these two countries as well as hands on knowledge which were important to my work in the Pacific region. I have always maintained that Technical Staff from the SPC and other regional organisations need this kind of exposure. Not only for networking, establishing contacts, but also for "first hand" experience which will become invaluable in their region.

CHAPTER 8. REGIONAL PLANT PROTECTION PROJECTS.

The South Pacific Commission Plant Protection Service (SPC-PPS) had a large number of regional projects between 1993 and 1996 (ref. SPC Plant Protection reports 1993-1996, Noumea, New Caledonia or SPC Library, Suva, Fiji) . More than $NZ20 million in budgets, it was one of the biggest investments that traditional donors have made in Plant Protection in the Pacific region. Most of the projects were initiated by Mr Robert Macfarlane, Plant Protection Officer and Co-ordinator for the SPC-PPS, who had done a magnificent job in the previous 6-7 (1987-1993) years or so. I had spent most of my time (1993-1996) completing those projects and ensuring their continuity.

There were 5 major projects on 1. fruitflies ($US1 million), 2. taro beetle ($SI 1 million), 3. plant protection ($F5million), 4. biological control ($F 1+ million) 5. plant protection in Micronesia ($AUD700,000).

In addition there was a joint project with the Commonwealth Agriculture Bureau International (Ref. CAB International HQ, London, UK or University of the South Pacific, Suva, PACINET headquarters) on

identification of the flora and fauna of the Pacific Islands ($US 8 million of which
$ US7.5 million was received according to the PACINET website). Several smaller projects were also initiated, including Taro Leaf Blight control.

Most of these projects were ongoing and have been for the 3 years prior to 1993. However, the funding were only received between 1993 and 1996, except the CAB International project which was approved later.

It was a lot of work to co-ordinate not only the project programmes but also the funding part. Most planning were done in meetings. For example, the CAB International project had to be discussed with member country counterparts in Plant Protection and the Environmental people. Two participants from each member country attended the meeting. In theory, there should be a representative from each of the 27 member countries making the total number of participants 54 plus 30 or so observers from other interested organisations.
CAB International spent $US100,000+ on that meeting alone with secretarial support from SPC-PPS. The $US 8 million budget was the result of group discussions by all member country counterparts, SPC-PPS and CAB International. It was a good result although there were objections to our project from SPREP (South Pacific

Regional Environmental Programme). I had explained to the SPREP representative to the formulation meeting that PACINET will actually help SPREP by identifying the flora and fauna they are trying to conserve and protect. She was of the opinion that PACINET will be in competition with SPREP for funds from their traditional donors.

CAB International representatives suggested that funds are available for all environmental projects from the Global Environment Facility which was established after the Rio Meeting and Organisations in the Pacific should support and complement each other as there is plenty of funds to go around. Needless to say, the $US7.5 million received for the project is proof of that statement.

The University of the South Pacific was chosen to implement the project after the funds were received, according to PACINET reports. This happened after I left, probably because the SPC-PPS were inundated with projects and planned work.

THE FRUITFLY PROJECT

This project was implemented by the Food and Agriculture Organisation (FAO) with management

support from the SPC-PPS. The expert Mr Allan Allwood worked with UNDP Volunteers and country counterparts in

Figure 52. Pawpaw is a promising export commodity from the Pacific Islands. Heat treatment developed by New Zealand is being tested here in the Cook Islands with assistance from SPC/FAO Fruitfly Project.

several Pacific countries including Tonga, Samoa, Fiji, Vanuatu, Solomon Islands, Federated States of Micronesia, Papua New Guinea and Cook Islands to study the fruitflies lifecycle and also look for a solution to this problem on a regional level. The project gradually extended activities to include all Pacific Island Countries. The picture (Figure 52) shows pawpaw harvests in the Cook Islands (Rarotonga), being prepared for heat treatment against fruit flies before export to New Zealand. Although it was a good arrangement, Cook Island pawpaw or pawpaw from other Pacific Islands are rarely seen in the Auckland supermarkets or fruitshops.

It is one of the obvious problems with small suppliers from the Pacific Islands. Availability in the market is not guaranteed, probably due to the low volume of production.

THE BIOLOGICAL CONTROL PROJECT

This project was funded by the German Government through its regional aid. There were experts in the field of biological control who worked with country counterparts on several sub-projects to control

Figure 53. A plot of cabbage where the Integrated Pest Management (IPM) SPC/German Biocontrol Project technologies were tested at the Sigatoka Valley, Fiji.

many pests and weeds in the region. These included *Mimosa invisa* and *Lantana camara* which were weed problems in the Pacific Islands. Insects included the integrated pest management (IPM) of diamond backed moth. *Encarsia ?haitiensis* was introduced to control the Spiralling Whitefly *Aleurodicus dispersus*, *Bracon spp* were also introduced to control coconut flatmoth in many Pacific Islands. The pictures show two of the

86

cabbage farms where the diamond back moth IPM programme was tested in the Sigatoka valley in Fiji.

Countries involved included Tonga, Samoa, Fiji, Cook Islands, Vanuatu, Solomon Islands and Papua New Guinea. The success of this project and Biological Control by NPPOs can be seen in many Pacific Island

Figure 54. IPM technologies help reduce dependence on poisonous chemicals used on cabbages through use of combined methods including biological control. This is one of the trial plots in the Sigatoka Valley, Fiji.

countries, to-day, where *Mimosa* and *Lantana* have disappeared from most agriculture areas.

Project Staff included Dr Jurgen Shaeffer (Teamleader), Dr Carlos Klein-Koch (Entomologist), consultants and several technicians and secretaries.

Note. The rhinoceros beetle was a huge problem in Tonga and many Pacific Islands when I was still in high school. Most coconut palms beside the roads had damaged leaves. Now, it is very rare to find coconut trees with damaged leaves due to biological control of the rhinoceros beetle.

PLANT PROTECTION IN THE PACIFIC

This project was funded by the European Union and implemented by the SPC-PPS. I was the Project

Manager of this project. A total of $F5 million was made available to the project through the European Union office in Fiji and the Forum Secretariat. It was part of the Lome Convention assistance to ACP (Africa, Caribbean, Pacific) countries.

The aim was to support training in all Plant Protection areas of concern in the Pacific, supply of equipment to National Plant Protection Organisation (NPPO) counterparts and staff funding.

One of the major activity of this project was to find a solution to the problem of Taro Leaf Blight

Figure 55. Taro Leaf Blight destroyed the Samoan Taro Industry in 1993. Extensive necrotic tissue on the taro leaf can be seen here. Most plantations are completely destroyed by the disease within a few weeks.

in Samoa. This disease destroyed Samoa's largest export in 1993. Two large meetings were held in Samoa and Papua New Guinea to bring together all the known experts throughout the world to make recommendations as to how we should solve the problem.

Figure 56. One of the TLB resistant taro plots at the Lae Research Station, Department of Agriculture, PNG.

There were many findings of the meetings, but the TLB resistant cultivars bred by Dr Anton Ivancic and DAF (Department of Agriculture and Forests) staff at Lae, PNG was the most obvious choice. They have successfully bred taro that were resistant to the disease and also taste good.

Figure 57. New TLB resistant taro being tested at the Lae Research Station, PNG.

I visited the Lae Research Station many times during the rainy and dry seasons to assess the resistance of their taro varieties and to join in the tasting panels.

A project was planned with the University of Technology in Queensland to clean the taro from PNG there before it is distributed to farmers in Samoa. The Australian Centre for International Agriculture Research (ACIAR) approved it with a proposed funding of $AUD 1 million, sadly it had to be cancelled when I left.

The SPC-EU Pacific Plant Protection Project also provided secretarial support to the other meetings as well as all SPC-PPS trainings of National Plant Protection Organisation (NPPO) counterparts. Equipment to be supplied to NPPOs was aimed at supporting and implementing the Global Plant Quarantine Standards recommendations from FAO/IPPC.

The project funds made many activities possible in the Pacific including trade negotiations between Tonga, Samoa and Fiji which was organised by the Forum Secretariat, Trade Officer, Mr Edgar Cocker. I had agreed to fund the travel and expenses of the NPPO officers of those countries, because I believe that trade between the islands is an excellent opportunity which was not often discussed in the past. The trade negotiations have led to many exchanges in produce like watermelon exports from Tonga to Samoa, for example.

The project also funded all SPC-PPS publications and Pest Alerts. All ACP countries were included in this project namely Tonga, Samoa, Fiji, Vanuatu, Solomon Islands, Kiribati, Tuvalu and Papua New Guinea). In addition, 3 French territories (New

Figure 58. An Extension/Training, Meeting/Workshop at the Koronivia Research Station, MAFF, Fiji.

Caledonia, French Polynesia and Wallis and Futuna) and 3 New Zealand managed countries/territories of Cook Islands, Niue and Tokelau. New Zealand provided counterpart funding for those 3 countries/territories. Staff of this project included myself, Semisi Pone as the Project Manager, the Biological Control Officer, Mr Albert Peters, a Training/Information Officer, , a Biocontrol Technician, Tissue Culture Assistant, and 1 Personal Assistant.

TARO BEETLE PROJECT

Taro beetle attacks and eat large holes on the taro corms in Vanuatu, Solomon Islands and Papua New Guinea. The project was aimed at studying the beetle life cycle and finding biological control and other control measures.

It was funded by the European Union through its Pacific Regional Agriculture Programme (PRAP) and managed by the SPC-PPS. It was renewed with $SI 1 million for another 3 years.

Staff of this project did an excellent job. Dr Brian Thistleton, Dr Billy Theunis and Mr Ioane Aloali'i carried out the work with assisting technicians.

Note. There are many problems of taro in the Solomon Islands where this project was based. Taro leaf blight, alomae and bobone viruses and taro corm beetle.

PLANT PROTECTION IN MICRONESIA.

This project was funded by Australian AID (AUSAID) by providing $AUD 700, 000 for activities in the 3 countries of the Republic of Palau, Federated States of Micronesia and the Republic of the Marshall Islands.

One expert was provided by the SPC-PPS to work in the 3 countries with NPPOs. Support in measures to implement the Global Plant Quarantine Standards and provide any required equipment were the main concerns of the project.

The Biosecurity Expert was based in the Federated States of Micronesia to implement this.

The SPC-PPS recruited Mr Dennis Kelly from the Australian Plant Quarantine Department to carry out this work.

Note: Plant Quarantine in Micronesia is crucial to the "Pacific Curtain" because of its direct links to Asia. There are many plant diseases and insect pests in Asia that have not been introduced to the Pacific Islands.

THE PACIFIC BIOSYSTEMATICS NETWORK (PACINET).

This project was proposed by CAB International as a way to find funds to identify all the fauna and flora of the Pacific Island countries. The SPC-PPS would implement the project with assistance from CAB International.

Figure 59. Participants at the Global Meeting on the Biosystematics Network, University of Cardiff, Wales. Author is on the left. Dr David Penman of New Zealand is third from left. Professor Tecwyn Jones, the organiser of the meeting is second from right.

A formulation workshop was held at the Tanoa Hotel in Nadi Fiji, in February 1996, where NPPOs and Pacific Environmentalists from member countries drafted the work programme and budget. A total of $US 8 million was requested from the Global Environmental Facility (GEF) for the project. It was one of the outcomes of the Global Environmental Meeting in Rio de Janeiro. According to PACINET reports $US 7.5 million was received for this project.

Professor Tecwyn Jones and his team from CAB International were the counterparts to the PACINET team in this project. The SPC-PPS acted as the secretariat for PACINET until the funds are approved and its own staff recruited. Activities of this Network can be obtained from reports available on the internet.

CAB International provided the central management for all the networks of this global project. PACINET only dealt with Pacific countries. Other networks were also established in other regions.

Note: **PACINET** was established to try and identify all fauna and flora in the Pacific Island countries. It would be nice to know which ones are important for various reasons including economics, conservation, biological control and so on.

CHAPTER 9. FUTURE PROSPECTS FOR BETTER BIOSECURITY IN THE REGION.

The Pacific Plant Protection Organisation (PPPO) was established to improve biosecurity in the Pacific region by improving linkages with other RPPOs and the United Nations Food and Agriculture Organisation. A Global Database was planned to link all RPPOs with Rome (FAO Headquarters) so reporting of new pest introductions and outbreaks can be reported at the speed of the internet. Faster action means border control can be established in a matter of days when there is a threat of a new pest being introduced into a susceptible country. Although there was a regional reporting system used by SPC, in 1994, called "Pest Alert" it was a bit slow and took weeks to alert all the countries around the region.

National Plant Protection Organisation counterparts meet every 2 years during the RTMPP (Regional Technical Meeting on Plant Protection) to discuss Plant Protection issues of importance to the region and also plan future activities of the SPC-PPS.
The RTMPP was linked to the RPPO technical meetings through the PPPO Executive which was the Head of the Plant Protection Service and Plant

Protection Advisor. The PPPO should play a major role in the biosecurity of the region in the future as it is the connection with the other RPPOs around the world and FAO, Rome.

The Global Database should also be linked to all SPC members. Pest reporting and border control will become easier as everyone are alerted to future dangers when pests start to move across from Asia, for example. The banana leaf roller and papaya ringspot virus are two examples. Cassava virus is another. They can really harm these important crops in the Pacific Islands and threaten food security in the region. The total destruction of taro in Samoa by taro leaf blight is a good reminder.

PLANT QUARANTINE STANDARDS

One of the important work that the SPC-PPS was involved in, for example, was the International Plant Protection Commission (IPPC) and Biosecurity experts work at FAO, Rome to draft and propose Global Plant Quarantine Standards to RPPOs and NPPOs. It was aimed at preventing the use of unjustified quarantine measures as trade barriers. The World Trade Organisation was being planned at that time (1993-1994) and it was a timely intervention. The WTO was launched in January 1995 under the Marrakech

Agreement to replace the General Tariffs and Trade (GATT) agreement which has been used since 1948.

The Global Plant Quarantine Standards will help when the WTO oversees any complaints or challenges about unfair trade barriers, using plant quarantine, by many countries.

The Pacific Islands are often disadvantaged when it comes to trade negotiations. They usually do not have a card up their sleeve, leaving them at the mercy of larger countries. Even if they have, it would be too expensive to challenge such measures at the WTO.

It was one of the concerns I raised at the Biosecurity Expert Meetings and RPPO Technical Consultations. One example is the FAO standard of "Prior Informed Consent" (PIC) in the case of Agricultural Chemicals. Small island nations do not have the expertise or the funds to check every agricultural chemical sold to them. The initiative lies with the supplier to inform them of any concerns regarding the chemical they are selling. For example, many agricultural chemicals are banned in some countries because of safety concerns, such as DDT. DDT was widely used because it was thought to be safe. Later it was found to contaminate just about everything in nature and accumulates in animal and human fat tissue, to be released when fats are broken down. Many farmer's deaths in Tonga from cancer/leukaemia were blamed on DDT. In one case he was a banana grower who used DDT powder to control scab moth with no safety gear whatsoever.

Information on safety issues must be made available to Pacific Island Countries, by the supplier countries, before any sale of such chemicals is made to them.

The same principle would apply with Plant Quarantine Standards in the sense that larger trading partners treat smaller nations fairly by informing them of their reasons for any Plant Quarantine concerns that may be viewed as "unfair barriers" to trade.

One example is the problem of fruitflies in Tongan watermelons. Watermelon export to New Zealand was a very good income earner for many farmers in Tonga. However, the discovery of fruitfly eggs in exported watermelons in Auckland in 1986, destroyed the industry. New Zealand MAFF and Tongan MAFF tried to solve the problem over many years by finding treatments that would eliminate fruitfly eggs from exported watermelons. It took 10 years, or so, before Tonga was able to export watermelons to New Zealand again. Mainly through help from New Zealand MAFF and others, to establish export protocols that satisfies New Zealand MAFF quarantine requirements.

The reason is not to prevent Tonga from exporting watermelons but to protect New Zealand's fruit industry, worth billions of dollars, from fruitflies that may destroy it. Both parties are satisfied with the protocols.

Negotiations should be made with plenty of goodwill on both sides, which is an excellent example of how the Global Standards in Plant Quarantine should work.

TRADE NEGOTIATIONS

One obvious advantage of the Global Plant Quarantine Standards is that all parties work on a level playing field. The same rules apply to everyone. It makes it easier to negotiate trade deals with neighbouring countries without having to setup quarantine measures and seek third party intervention, for example "experts", in every case. The methods of achieving the results may be different, for example the watermelon protocols, but the results are all the same, by using the Global Plant Quarantine Standards to facilitate trade.

This was proven to be true in the case of Tonga, Samoa and Fiji who took the initiative to negotiate their own trade deals between their respective countries in 1995, with assistance from the SPC-PPS and the Forum Secretariat. I believe in this case there was a "blanket" deal which covers almost all agricultural goods to be traded. Such agreements should be reviewed every 3 years or so to update all information regarding pest movements.

Pacific Island Countries (PIC) should look more to their neighbours for trade as many PICs now produce many agricultural products they did not produce before. They also make a lot more money than in the past. It may improve total country income, for all concerned, amongst other benefits.

Note: Trade in the past was based on finding a buyer in overseas markets. Now any Pacific Islander can export and sell his own products anywhere in the world, if he has the resources to do so.

CHAPTER 10. REGIONAL TRADE

I would like to comment on economic trade in the Pacific Region especially exports of agricultural produce from the Pacific Islands to New Zealand and Australia. All of the work in Plant Protection in the Pacific region during my time there (1985-1996) was to support and enhance export of agricultural produce from the Pacific Islands to metropolitan centres like Auckland and Sydney where large populations of expatriate islanders live.

Rootcrops like yams, taro (*Colocasia* sp.), dasheen (*Xanthosoma* sp.) and cassava are important staples in most of the Pacific Islands. During the last 50 years, a large number of islanders shifted to New Zealand, Australia and the United States. They also take their foods, culture, language and oddities with them.

In New Zealand there are probably 200,000 to 300,000 Pacific islanders living there. It is a huge market for traditional growers of rootcrops from Tonga, Fiji and Samoa. It is estimated that the market for rootcrops in New Zealand and Australia may be more than $NZ 50 million per year and growing.

Unfortunately, similar to bananas, rootcrops are also grown in other countries. The Pacific Islands, probably lost more than $1 billion since the New Zealand green and ripe banana market was taken over by Ecuador and the Philippines in the 1980s to 1990s.

India and Vietnam have entered the New Zealand market and are supplying taro, yams and cassava in some shops in Auckland. If the Pacific Islanders don't get their act together, the rootcrop market will be lost to Asia. In 20 years, it will be another $1 billion gone to the competition.

If Pacific Islanders are no longer able to compete in the metropolitan markets with large countries such as Vietnam and India, but "pushed aside" every time, rootcrops will become another "banana episode" in the sad trade story of agricultural trade in the Pacific.

The current supplier of taro and cassava to the New Zealand market, and Australia, is Fiji. The taro leaf blight had wiped out the Samoan taro industry and gave Fijian growers an opportunity to increase their market share. The Fijians are doing a good job, but are probably not supplying enough, thus giving the Vietnamese and Indians a foot in the door.

Tonga is the current supplier of yams, giant taro and some dasheen, but unlike the Fijians, supply is not reliable, packaging is sub-standard and the whole establishment is "ad hoc" with no professional advisors to help growers who wish to export.

Other produce from the Pacific Islands also suffer from similar problems. Pawpaw from the Cook Islands and Fiji cannot supply the demand in Auckland. The large supermarket chains have no choice but to import from the Philippines and so on.

I spend more than one hour discussing these issues, 2 years ago, with the Deputy Directors of Agriculture in Tonga. I noticed they have made some improvements but it is not fast enough. They really have to step up or the market will be very crowded by the time they get organised. They have been doing it for more than 30 years and should be leading the market instead of just "surviving".

Rootcrops from India and Vietnam are professionally packaged and well presented, similar to bananas from Ecuador! I have tried their yams and cassava and they are excellent. They are the "team to beat" now. The Pacific Islanders will be playing "catch-up" soon if they do nothing.

The Pacific Islanders should respond to this threat to their traditional market or they will be gone out the door. India and Vietnam can supply all the rootcrops the Australian and New Zealand markets want and more. Similar to what Ecuador is doing with green and ripe bananas in New Zealand.

Although there are large numbers of Agriculture produce that can be exported from the Pacific Islands, they all suffer from the same problems of unreliable supply due to low production. Despite the millions of dollars poured into regional Plant Protection, it may be just a waste of money if islanders cannot turn those advantages into foreign exchange to boost their local economies.

All the IPM technologies, disease resistant and high producing varieties, production techniques, plant quarantine standards and other Plant Protection improvements count for nothing if they are not used to improve the livelihood of the people of the Pacific. It is the whole purpose of spending millions of dollars on Plant Production and Plant Protection projects in the region, not counting the decades spent in training the agriculture staff to carry out that work.

A lot of authors suggest the easy lifestyle and traditional values of the Pacific Islanders as excuses for not gearing their economies to 21th century production and export techniques. It may be true because as much as $NZ 800 million, or more, a year is poured into the islands from their relatives overseas in the form of remittance.

Tonga probably receives $NZ 300 million in remittance from New Zealand, Australia and USA every year. This includes cash, shopping and goods. Tongans overseas normally send cash to relatives for school fees, church activities, goods and so on. They also buy them shopping (eg. through the Melie-mei-Langi stores). They pay in New Zealand, Australia or USA and relatives pickup the shopping from the Melie-mei-Langi stores in Tonga. Vehicles, building materials, clothes and so on are also sent from overseas relatives.

Aid is also another factor. In Tonga, 'Akilisi Pohiva, people's representative to the Tongan Parliament, suggest that 51% of the GDP (about $TOP799 million, Statistics Department, Tonga) is aid money. That is a whopping $TOP 407, 490, 000 per year in aid!

If you add up the remittance and the AID, it could be as much as 80% of the Gross Domestic Product! (GDP). They are "easy money", thus maintaining and promoting the "easy lifestyle" in the islands often quoted by many authors. It may not be all bad as it gives the islanders an "economic cushion" to prop them up while they develop their exports, tourism and industries. Hopefully, when the "crunch" comes they will be ready. That is, when AID and remittance money falls below $NZ 100 million, for example.

This is how Fijian taro (Figure 60) is sold in the shops

in Auckland. They are normally sold fresh. The Fijians have perfected the technique of exporting fresh taro to New Zealand. They are clean, fresh with no rotting or physical damage! At least, 2 varieties are available. The pink and white taro.

Figure 60. Fresh taro being sold in the small shops in Auckland, New Zealand.

This is how Tongan yam is being sold in the Auckland shops (Figure 61). They are normally packed in 3 kg, unprinted, food grade? plastic bags

Figure 61. Frozen Tongan yams being sold in the small shops in Auckland, New Zealand.

and frozen. Although the product itself (yam) is excellent. The

packaging needs a lot of improvement. The exporter needs to look at how the Vietnamese are packaging their yams and steal a few ideas from them.

Figure 62. Fijian frozen cassava on display in a small shop in Auckland.

This is how Fijian cassava is sold in the small shops in Auckland (Figure 62). Fijian Cassava is the most dominant in the Auckland market. The product is excellent and the

Figure 63. Local vegetables sold in the small shops in Auckland, New Zealand. Most vegetables are produced in glasshouses during the winter months.

packaging is professionally done. Supplying vegetables during the winter months was a good income earner for many growers in the Pacific Islands but New Zealand farmers are now producing them in glasshouses in the winter months. These are local

108

vegetables (Figure 63) in a small shop in Auckland. They have excellent quality and prices are low compared to those supplied from the Pacific Islands. There are no rootcrops currently sold in the 2 large supermarket chains in New Zealand. They are Progressive Enterprises Ltd (Foodtown and Countdown Stores) and Food Stuffs Ltd (Pak n Save, New World Stores).

Presentations of fruits and vegetables at the supermarkets are excellent and probably conducive to sales. They appear fresh and crisp with good colour. Although presentations

Figure 64. Presentation and quality is everything at the supermarkets in Auckland. This is why produce from the Pacific Islands cannot compete with locally produced fruits and vegetables.

in the smaller shops are not as ordered, they do sell a lot of the rootcrops from the Pacific and must be considered in any marketing plans for the future. Both frozen and fresh rootcrops should be supplied especially in the case of taro and yams. There are many agriculture and other products that can be imported from the Pacific Islands but they are not! Many of these products can now be seen in the Asian supermarkets and stores. They are newcomers to the Auckland market and they don't waste any time! The countries of the Pacific can still supply their products to their nearest markets in Australia, New Zealand and USA, but they need to do everything themselves. Instead of waiting for somebody else to

come along and buy their products, they should package them, setup their own shops anywhere in the world and sell their products! No need to look for a market! They should

Figure 65. Taro and xanthosoma are popular rootcrops for Pacific Islanders in Auckland.

steal some ideas from the Asian Supermarkets and stores in Auckland. They sell everything from imported frozen orange leaves to green coconuts, taro, yams, frozen cassava, fruits and so on. If they can sell it, it's already packaged and in the fridge, freezer or the shelves.

Many produce such as breadfruits, seafood, and leafy vegetables such as "pele" and so on are popular with Pacific Islanders in Auckland. They can still supply these produce, and others, from the islands but they need to hurry up!

FURTHER READING...

1. Clark, M.F. and Adams, A.N. (1977). Characteristics of the Microplate Method of Enzyme Linked Immuno-Sorbent Assay for the detection of plant viruses. Journal of General Virology. 34,475-483.

2. Davis, R.I., Brown J.F., and Pone S.P. (1996). Causal Relationships between Cucumber Mosaic Cucumovirus and Kava Dieback in the South Pacific. Plant Disease. 80:194-198.

3. Dingley, JM., Fullerton, R.A. and Mckenzie, E.H.C. (1981). SPEC/UNDP/FAO Survey of Agricultural Pests and Diseases. Volume 2. (Cook Islands, Fiji, Kiribati, Niue, Tonga, Tuvalu, Western Samoa), 485 pp.

4. Elliot, J.M. (1977). Some methods for the statistical analysis of benthic invertebrates. Second Edition. Freshwater Biological Association Scientific Publications. No. 25, United Kingdom.

5. Food and Agriculture Organisation (United Nations) Biosecurity (Plant Quarantine) and RPPO Reports. Rome, Italy. 1993-1996.

6. Journal Indonesian SS AAS Vol.15, No.15, No.2:34-41.

7. Madden, L.V. and Campbell, C.L. (1986). Descriptions of virus disease epidemics in time and space. Pp 273-293. In Plant Virus Epidemics, Monitoring, Modelling and predicting outbreaks. Eds. Mclean, G.D., Garret, R.G. and Ruesink, W.G. Academic Press, Sydney, pp 550.

8. Mossop, D.W. and Fry, P.R. (1984). SPEC/UNDP/FAO Survey of Agricultural Pests and Diseases; Records of viruses pathogenic on plants in the Cook Islands, Fiji, Kiribati, Niue, Tonga and Western Samoa, Volume 7. 9 pp.

9. Pearson, M.N. and Pone, S.P. (1988). Viruses of vanilla in the Kingdom of Tonga. Australasian Plant Pathology 17(3), 59-60.

10. Pearson, M.N., Brunt, A.A. and Pone, S.P. (1990). Some hosts and properties of a potyvirus from *Vanilla fragrans* in the Kingdom of Tonga. Journal of Phytopathology 128:46-54

11. Plant Protection Service Reports. South Pacific Commission (Now Secretariat for the Pacific Community). Suva, Fiji & Noumea, New Caledonia. 1993-1996.

12. Pone, S.P. (1988). An investigation of 3 virus diseases of *Vanilla fragrans* (Salisb.) Ames in the Kingdom of Tonga. MSc Thesis (Available at Auckland University Biological Science Library and the National Library of New Zealand).

13. Pone, S.P. Epidemiology of Vanilla Necrosis Potyvirus in *Vanilla fragrans* (Salisb.) Ames plantations in the Kingdom of Tonga (2013). Rainbow Enterprises, Auckland.

14. Pone, S.P. Development of an ELISA Tests for Vanilla Necrosis Potyvirus in *Vanilla fragrans* (Salisb.) Ames in the Kingdom of Tonga (2013). Rainbow Enterprises, Auckland.

15. Pone, S.P. and Pearson, M.N. (1988). Koe mahaki vailasi 'o e vanilla. Virus diseases of vanilla in Tonga. Advisory Leaflet. MAFF, Tonga. 8pp.

16. Research Division Reports, MAFF, Tonga. 1985-1992.

17. Tissue Culture Reports. Institute for Research, Extension and Training in Agriculture, University of the South Pacific, Alafua Campus, Samoa . 1992-1993.

18. Van der Plank (1960). Analysis of epidemics, pp 229-289. In Plant Pathology, Volume III. Eds. Horsfall, J.G. and Dimond, A.E. Academic Press, New York, 675 pp.

19. Wisler, G.C., Zettler, F.W. and Mu, L. (1987b). Virus infections of Vanilla and other orchids in French Polynesia. American Orchid Society Bulletin 56 (4), 1987.

www.ingramcontent.com/pod-product-compliance
Lightning Source LLC
Chambersburg PA
CBHW071801200526
45167CB00017B/801